The Future of Mobile Communications

Also by Peter Curwen

RESTRUCTURING TELECOMMUNICATIONS: A STUDY OF EUROPE IN A GLOBAL CONTEXT

UNDERSTANDING THE UK ECONOMY (4th ed)

INDUSTRIAL ECONOMICS (with R. Stead & K. Lawler)

PRINCIPLES OF MICROECONOMICS (with P. Else)

The Future of Mobile Communications

Awaiting the Third Generation

Peter Curwen

palgrave
macmillan

First published 2002 by
PALGRAVE MACMILLAN
Houndmills, Basingstoke, Hampshire RG21 6XS and
175 Fifth Avenue, New York, N.Y. 10010
Companies and representatives throughout the world

PALGRAVE MACMILLAN is the global academic imprint of the Palgrave Macmillan division of St. Martin's Press, LLC and of Palgrave Macmillan Ltd. Macmillan® is a registered trademark in the United States, United Kingdom and other countries. Palgrave is a registered trademark in the European Union and other countries.

ISBN 1–40390–268–2

This book is printed on paper suitable for recycling and made from fully managed and sustained forest sources.

A catalogue record for this book is available from the British Library.

Library of Congress Cataloging-in-Publication Data
Curwen, Peter J.
The future of mobile communications : awaiting the third generation / Peter Curwen.
p. cm.
Includes bibliographical references and index.
ISBN 1–40390–268–2 (cloth)
1. Telephone, Wireless. 2. Wireless communication systems. I. Title.

HE9713 .C87 2002
384.5′3—dc21 2002072509

10 9 8 7 6 5 4 3 2 1
11 10 09 08 07 06 05 04 03 02

Printed and bound in Great Britain by
Antony Rowe Limited, Chippenham, Wiltshire

This book is dedicated to the memory of
Montague Curwen 1913–2001

Contents

List of Figures and Tables

Figures

Tables

Glossary

Analogue: A method of modulating radio signals so that they can carry information in the form of voice and data.

Asymmetric digital subscriber line (ADSL): A digital transmission technique for enhancing the information-carrying capacity of copper cables in the local loop via signal compression. It is designed for video-on-demand services by delivering video signals to customers, at the same time as voice signals, with a low-capacity return channel for subscriber ordering purposes.

Bandwidth: The quantity of spectrum required for a specific purpose. An analogue telephone call normally occupies 4 KHz (kilohertz), where Hertz is a measure of cycles per second. Digital systems are measured in binary bits per second (bps).

Base station: A central radio transmitter/receiver that connects to mobile devices within a given range.

Bluetooth: An initiative set up to create a short-range wideband radio standard to allow mobile phones, palmtops and portable PCs to communicate with each other without cables.

Broadband: Broadband networks can convey large volumes of information, and hence require a capacity of at least 4 MHz in respect of an analogue network and 2 Mbits (megabits – million binary digits or 'bits') per second in respect of a digital network. Networks with a modest capacity of 4 KHz (analogue) or 64 Kbits per second (digital) are known as narrowband or, in the intermediate bandwidth, wideband.

Byte: Eight bits of data.

cdma2000 1xRTT: A CDMA standard with data speeds equivalent to GPRS and hence best treated as 2.5G. However, unlike GPRS, which is a pure data companion for GSM, cdma2000 1xRTT has the ability to increase voice capacity by 1.5–2 times.

cdma2000 3xRTT: A CDMA standard providing up to 2 Mbps.

Cellular systems: Mobile radiocommunications networks in which an overall large area of service is divided into a number of cells each having its own low-powered transmitter/receiver equipment (base station). The use of cells allows the same frequency to be re-used in different cells. As subscribers move from one cell to another, the cellular system automatically hands on the call to the base station in the next cell.

Digital: The coded representation of a waveform by binary digits which takes the form of pulses of light in a fibre-optic network. An analogue system represents the waveform *per se*.

Digital enhanced cordless telecommunications (DECT): Digital cordless standard developed by ETSI, supported by Directive 91/288/EEC which establishes harmonised frequency bands for DECT. DECT provides an alternative to the local loop access to the public switched telephone network.

Dual band: A feature on certain mobile handsets which allows them to use two frequency bands.

Dual mode: A feature on certain mobile handsets which allows them to handle both analogue and digital signals.

European telecommunications standards institute (ETSI): The European standards organisation in the telecoms field.

Extensible mark-up language (XML): A standard that will make the resources of the World Wide Web available over the phone.

Fibre optics: In 1980 a single fibre was capable of operating at 45 Mbps (million bits per second). The current maximum speed is roughly 10 Gbps (billion bits per second) and the target speed for 2010 is roughly 100 Gbps.

Gateway: A facility that adapts signals and messages of one network to the protocols and conventions of other networks or services.

General packet radio service (GPRS): A packet-switching technique that takes wireless communications into the datacoms arena, allowing users to stay permanently on-line. It gives users increased data rates up to 115 Kbps. It is usually an extension of an existing GSM network and widely known as 2.5G.

Global system for mobile communications (GSM): Strictly *Groupe Speciale Mobile*, the central standard developed by ETSI for digital mobile systems, using TDMA (time division multiple access) techniques. Harmonised frequency bands for GSM are supported by Directive 87/372/EEC. System features include roaming (see below). The user receiving a GSM call pays for the segment outside the caller's own territory.

High-speed circuit switched data (HSCSD): A circuit-switched technology able to operate at relatively high speeds of up to 57.6 Kbps.

Incumbent (European Union): Telecommunication organisation(s) granted special and exclusive rights by member states (Commission Directive 90/388/EC of 28 June 1990) or public operator(s) which enjoyed a *de facto* monopoly before liberalisation. In the case of mobile telephony, the word 'incumbent' may refer to the subsidiary of the incumbent in the fixed-wire telephony market.

Instant messaging service: This enables Internet users to view a 'buddy list' installed via software on their PCs, and to send messages to any of them who are connected at the time. Whatever is typed appears instantly on the recipient's screen.

Integrated services digital network (ISDN): A set of network standards which are designed to permit an existing network to expand its capacity and functions via digital technology.

International mobile telecommunications 2000 (IMT-2000): The term used by the International Telecommunication Union, a United Nations Agency, to describe 3G. It encompasses a number of mobile telephony standards that meet certain requirements in terms of transmission speed and other matters to provide the capability of delivering mobile multimedia services.

Internet: A global network of networks, accessed via a computer and a modem.

Internet portal: A site which is initially accessed when making an Internet connection, containing search engines, news, e-mail and other facilities.

Java: A computer language for use on the Internet developed by Sun Microsystems. It is, for now, the *de facto* standard for creating Web applications.

Local area network: A data communications network which is limited in area – typically to a 1 km radius – thereby permitting optimisation of the network signal protocols and ensuring speedy data transmission.

Mark-up language: This presents data or content by attaching a series of presentation instructions called tags. The tags 'mark up' the content such that it can be identified and formatted.

Mobile virtual network operator (MVNO): Strictly an operator owning a mobile switching centre, possessing a unique mobile network code and issuing its own SIM cards while providing services over another operator's network.

Modem: The modulator/demodulator is a device that converts digital signals into analogue format for transmission via analogue networks.

Packet-switched data services: Packet- and circuit-switched data services means the commercial provision for the public of direct transport of data between public switched network termination points, enabling any user to use equipment connected to such a point in order to communicate with another point.

Personal communications networks (PCNs) or services (PCS): PCNs were introduced by new entrants to the mobile phone market using the DCS-1800 standard developed by ETSI. Compared to GSM, PCN cells are smaller and work at higher frequencies. They are designed to work with small, lightweight, low-power handsets. In the USA they are known as personal communications services and operate at 1900 MHz.

Personal digital cellular (PDC): A Japanese standard for digital mobile telephony in the 800 MHz and 1500 MHz bands.

Personal handyphone system (PHS): A Japanese standard for digital mobile telephony in the 1900 MHz band.

Portal: A site initially accessed by Internet users that offers a range of services including news, Internet search, e-mail and financial and other information.

Protocol: A set of rules and formats that governs the exchange of information between two or more entities.

Roaming: Whereby a subscriber to one network can use radio telephone equipment on any other network that has entered into a roaming agreement in respect of the same or other countries for both outgoing and incoming calls.

Short message system (SMS): A variant of instant messaging (see above) for mobile devices.

Spectrum: The complete range of electromagnetic frequencies.

Subscriber identity module (SIM) card: A small printed circuit board that is inserted into a handset to instigate service provision.

Synchronous: A data stream with the same capacity in both directions.

Time division(al) multiplexing: A technology which increases the speed with which signals pass along a fibre-optic cable. The current optimum speed is 2.5 Gbps.

Very small aperture terminals (VSAT): Very small aperture terminals permit the reliable transmission of data by satellite, using a comparatively small antenna of three to six feet. This can be attached to existing terminal equipment, operating like an aerial modem.

Virtual network operator (VNO): The simplest variant of a VNO involves branding and billing arrangements provided by a retailer (or independent service provider) which is reselling an existing network's offerings. However, a more sophisticated VNO can set up its own tariffs and special services that are not available from the owner of the network whose capacity is being leased, and from the customer's viewpoint appears to be a wholly separate supplier.

Wave division(al) multiplexing (WDM): Instead of sending a light signal of a single colour down a fibre-optic cable, WDM uses many colours simultaneously, each carrying a separate stream of information. A currently available 40-channel fibre can carry 1.3 million simultaneous conversations. Also known as DWDM.

Wideband code division multiple access (W-CDMA): A third-generation mobile services platform based on a protocol structure similar to GSM.

Wireless application protocol (WAP): A global open data protocol that allows users to access online services from their small-screen mobile phones using a built-in browser. It was standardised at the WAP Forum (www.wapforum.org) established in June 1997. Specifications for WAP-Ng, the successor to WAP1.X, were released in August 2001. These contain a link-up with i-mode (see above).

Abbreviations

1G	Original analogue technology
2G	Second-generation mobile telephony
2.5G	An intermediate mobile technology between 2G and 3G
3G	Third-generation mobile telephony
3GPP	Third generation partnership project
ADSL	Asymmetric digital subscriber line
AMPS	Advanced mobile phone service
ARPU	Average revenue per user
B-GAN	Broadband global area network
BPMB	Bundled price per megabyte
BREW	Binary runtime environment for wireless
CATT	Chinese Academy of Telecommunications Technology
CDMA	Code division multiple access
cHTML	Compact hypertext mark-up language
DECT	Digital enhanced cordless telecommunications
DS-FDD	Direct sequence frequency division duplex
EBITDA	Earnings before interest, tax, depreciation & amortisation
EDGE	Enhanced data [rates] for GSM evolution
ERC	European Radiocommunications Committee
ETSI	European Telecommunications Standards Institute
EU	European Union
EV-DO	Evolution-data only
EV-DV	Evolution-data and voice
FCC	Federal Communications Commission
FDD	Frequency division duplex
FOMA	Freedom of mobile multimedia access
GEO	Geostationary earth orbit
GHz	Gigahertz (billion hertz)
GMPCS	Global mobile personal communications by satellite
GPRS	General packet radio service
GPS	Global positioning satellite
GSM	Global system for mobile
HDML	Hand-held device mark-up language
HSCSD	High-speed circuit switched data
HTML	Hypertext mark-up language
IMT-2000	International mobile telecommunications-2000

IP	Internet protocol
ITU	International Telecommunication Union
J2ME	Mobile Java
Kbps	Kilobits (thousand bits) per second
LAN	Local area network
LEO	Low earth orbit
MAN	Metropolitan area network
MAP	Multi access portal
Mbps	Megabits (million bits) per second
MC-FDD	Multi-carrier frequency division duplex
MEO	Medium earth orbit
MHz	Megahertz (million hertz)
MITA	Mobile Internet technical architecture
MMDS	Multichannel multipoint distribution services
MMS	Multimedia messaging service
MVNO	Mobile 'virtual' network operator
NGSO FSS	Non-geostationary fixed satellite service
NMT	Nordic mobile telephone
NTIA	National Telecommunications and Information Administration
OS	Operating system
PC	Personal computer
PCN	Personal communications network
PCS	Personal communications services
PDA	Personal digital assistant
PDC	Personal digital cellular
PHS	Personal handyphone system
PIM	Personal information manager
RAN	Radio access network
RBOC	Regional Bell operating company
RF	Radio frequency
RTT	Radio transmission technology
SABA	Simultaneous ascending bid auction
SAC	Subscriber acquisition cost
SDMA	Space division multiple access
SIM	Subscriber identity module
SMS	Short message service
TDD	Time division duplex
TDMA	Time division multiple access
TD-SCDMA	Time division synchronous code division multiple access

TIM	Telecom Italia Mobile
TMT	Technology, media & telecommunications
UMTS	Universal mobile telecommunications system
UTRA	Universal terrestrial radio access
VSAT	Very small aperture terminal
WAP	Wireless application protocol
W-CDMA	Wideband code division multiple access
Wi-Fi	Wireless fidelity
W-LAN	Wireless local area network
WML	Wireless mark-up language
WRC	World Radiocommunications Conference
xHTML	eXtensible HTML
XML	eXtensible mark-up language

Acknowledgement

During the past several years that this project has been in gestation, I have published a series of articles in *info – the journal of policy, regulation and strategy for telecommunications, information and media*. Some of the ideas behind the material in this book were first developed in *info*, and I would like to take the opportunity to thank the editor, Colin Blackman, for his support both in relation to these earlier efforts and their development into this book.

Preface

The aim of this book is to investigate the prospects for mobile communications in the new millennium. In effect, this boils down to a key issue: Will the so-called third generation of mobile (alternatively, wireless or cellular) technology turn out to be a success or a failure? This is no minor issue since telecommunications is one of the world's leading industries, and one that is increasingly moving from a fixed-wire to a wireless basis. Furthermore, whereas widespread fixed-wire telecommunications have historically been the preserve of developed countries, the advent of mobile telephony has made it possible for less-developed countries to improve their communications capability very substantially and speedily without the need to undertake impossibly costly investment in fixed-wire links. Thus it is China that is currently the largest market for basic mobile telephony in the world, having recently overtaken the USA.

However, there is a clear difference between the basic voice telephony-driven networks that are currently in common usage, and the data-driven networks that are gradually being introduced. No one questions that the level of demand for basic voice telephony, however transmitted, will continue to grow, but the underlying difficulty is that the explosion of demand for wireless-delivered voice telephony has already created a situation whereby networks are close to saturation point in developed markets. Hence, network operators need to introduce new, high-value-added services if they are not to see their average revenue per user (ARPU) decline as they seek out the remaining marginal customers who neither want to spend much at all on calls nor to buy expensive, unsubsidised handsets. The trouble is that the basic technology and networks developed for voice are inadequate for more sophisticated purposes, although it does have to be said that the development of the short message service (SMS) has proved to be an unprecedented and, in truth, largely unanticipated success based on existing technology.

In principle, the most critical constraint on what services can be delivered via a mobile device of some kind – which may not come in the form of a conventional handset – is related to the speed at which the data can be transferred. Large data files are necessarily expensive to transmit at slow speeds, and, in any case, potential customers are becoming increasingly impatient as they get used to the power of their office and, where applicable, home personal computers, to deliver large volumes of almost instantaneous information. Developing new technology to permit the transfer of large volumes of data is, unfortunately, very expensive, and handsets present particular problems because more sophistication means more embedded technology, yet customers expect the handsets themselves to become progressively smaller and lighter rather than the reverse. In addition, there is the thorny issue of licences. One of the classic things that are by their very nature in short supply is radio spectrum. Since spectrum must of necessity be allocated to serve only those practical purposes known at the time of its initial allocation, it is not typically held in reserve for theoretical purposes which may or may not appear at some point in the future. When new purposes do appear, therefore, it may be hugely difficult to make sufficient space for them in the parts of the spectrum range best suited for the purposes in question.

Clearly, therefore, new services requiring much more bandwidth than voice present extremely difficult choices on the supply side of the market if they are to be delivered via wireless. Equipment manufacturers must be willing to invest huge amounts of money to develop new or upgraded technologies, and network operators must invest heavily to build new networks and to acquire the spectrum to enable the services to be transmitted – all before any revenue rolls in. Traditionally, all areas of the communications industry, not solely telephony, have tended to operate on the historically sound principle that 'if you supply it, they will come'. This assumption was applied in the initial development phase of mobile telephony as it graduated from an analogue to a digital technology, and yet again was shown to be sound. Not surprisingly, therefore, it was also applied when it became time to shift the emphasis significantly from mobile voice to mobile data – only this time things did not turn out quite so happily, at least initially, because whereas it is easy enough to envisage why everyone wants to communicate, it is much less easy to determine what else they need sufficiently badly to be willing to pay an economic price for it. This is not to say that the story will lack a happy ending, rather that whatever ending is forthcoming will be at the end of a very rocky road, and one that is probably going to be strewn with casualties, so the story provides a fascinating case study for students of business and management.

The story of the evolution of what is always now known simply as 3G forms the central subject matter of this book. It is a story that has been unravelling over a period of several years, and has necessitated a revision of this manuscript on virtually a daily basis. However, the point has now been reached where it is possible not merely to recount the events so far, but to identify every issue that has arisen and subject each to some analysis in order to form a view as to the prospects for 3G. It is not easy – indeed, it is impossible – to find this story recounted as a single entity elsewhere in the public domain. Not surprisingly, although the issues are of considerable importance to industry players, the information tends to be set out for them in report form, and they are expected to pay huge fees to consultancies for the information which is then necessarily withheld from the public gaze.

A minor disclaimer or two is in order. First, every fact and number in this book has been cross-checked and they were all believed to be correct at the time of submitting the manuscript in April 2002. However, at various points the author has been obliged to contact the media, websites and consultancies to query published data which differed from his own records, so it has to be accepted that, where 3G is concerned, there is a considerable amount of misleading data in circulation, in part because the industry is evolving at an unprecedented rate – but hopefully very little in this book. The author would nevertheless be only too pleased to be told of any factual errors remaining other than those arising because time has moved on.

Second, there is a surprising amount of discrepancy in the naming of organisations/technologies. It is possible to argue that this should be viewed as a fairly trivial issue – for example, does CATT stand for Chinese Academy of Telecommunications Technology as in this text or China Academy of Telecommunications Technologies? Does the 'M' in SMS stand for 'message' or 'messaging'? Ultimately, there is no choice but to plump for the version most commonly cited in the media and advise readers that they may occasionally come across other versions.

Structure

The core chapters of this book are concerned expressly with third-generation mobile telephony (3G). However, before reaching this material there is a need to introduce information about the preceding history of mobile telephony covering what is known as 2G and 2.5G. Chapter 1 initially covers issues concerned with the 2G networks that are now commonplace in most parts of the world before moving on to introduce 3G and the technology that it involves. The intermediate step, 2.5G, is currently being introduced, and as it can only be dealt with in passing in Chapter 1, it forms the main part of Chapter 2 where several variants, the Japanese i-mode, the wireless application protocol (WAP) and the general packet radio service (GPRS) are introduced in more detail. There is also an extended discussion of the role of satellite, seeking among other things to explain why it is expected to play only a minor role in the move towards 3G.

Chapters 3–5 cover 3G in detail. Chapter 3 concentrates upon the licensing process in Western Europe because that is where most progress has been made so far. Chapter 4 then covers the rest of the world. These descriptive chapters underpin Chapter 5 which is the core chapter of the book in that it sets out to analyse in some detail every issue that is raised in the two previous chapters. Particular emphasis is placed upon the methodology of licence issuance – the so-called auction versus beauty contest debate. In the concluding section, an attempt is made to provide a realistic assessment of the prospects for 3G.

The final chapter is by way of a detailed case study looking at three companies with mobile networks and brands that are reasonably worldwide in their scope, although it has to be said that Vodafone meets that description rather better than Orange given the latter's lack of presence in the USA and Japan, and rather better than T-Mobile given its lack of presence in Japan and much of Western Europe, and hence is the main focus of the narrative. It is worth noting that Vodafone is a pure mobile operator, whereas the other operators heavily involved in 3G are, even if separately quoted, almost all linked to incumbent fixed-wire operators. Clearly, one of the effects of 3G will be a restructuring of the telecommunications industry, and the case study should prove useful in providing an underpinning to the wider debate as to how such a restructuring will occur.

1
Introduction to Mobile Telephony

Introduction

What follows is concerned with the evolution of, and prospects for, so-called third-generation or 3G services provided using mobile devices. It is particularly concerned with developments in Western Europe because that is where most progress has so far been made in terms of the number of countries involved, but full account is also taken of developments elsewhere in the world.

The mobile – alternatively wireless or cellular radio – market encompasses radiopaging and a variety of radio telephone services which are provided by connecting together two or more mobile devices or terminals via a base station. For the purposes of the discussion below these devices will be taken to be handsets since they dominate current usage, although the advent of 3G is going to lead to a considerable increase in diversity even if the dominant device will necessarily remain the handset in its current configuration. The issue of what form other devices might take is addressed at the end of Chapter 2.

Mobile communications have been commercially available in some form or other since 1946, although the initial format in the USA involved a single transmitter covering an entire city. Such a network had very little capacity and delivered calls via a fixed-wire link to the home. Mobile networks only truly developed once a cellular structure was adopted. Modern cellular networks commenced services in the USA in 1983. Interestingly, although the Nordic mobile telephone analogue system (NMT-450 operating at 450 MHz – short for megahertz or million hertz) was first introduced in Scandinavia as early as 1981, it was not adopted throughout Europe and there were multiple incompatible standards in use by 1985, at which point the Federal Communications

Commission (FCC), the main regulatory body for telecommunications in the USA, was imposing a common analogue advanced mobile phone service (AMPS) technology in the 800 MHz band. However, this situation completely reversed itself when digital networks became feasible. For its part, the US government took the view that competition would determine which standard was best, whereas the European Union (EU) set in hand, commencing as early as 1982, a process that would produce a single EU-wide digital standard.

Throughout the world the standard digital technology is currently known as second-generation or 2G. The original, analogue, technology (1G) is still to be found, quite extensively in the USA but even in parts of the EU where it is being phased out and the spectrum transferred to 2G networks which, because they are digital, use the spectrum much more efficiently. As shown in Table 1.1, most of the 2G licences in Western Europe were issued between 1992 and 1996, and utilised spectrum in the 900 MHz band known as the global system for mobile (GSM). The term GSM is often also used in relation to spectrum in the 1800 MHz band (the 1.8 GHz (gigahertz) band since 1 GHz is equivalent to 1000 MHz), although it is more usual to refer to this as either DCS 1800 or as a personal communications network (PCN). PCN spectrum was typically licensed in Western Europe after 1997 – either to existing GSM licensees to ameliorate spectrum shortages as mobile telephony surged in popularity or to permit the entry of new operators. A crucial factor was that the administrative arm of the European Union, the European Commission, enforced the exclusive usage of these two spectrum bands via the auspices of the European Telecommunications Standards Institute (ETSI).

Elsewhere in the world a variety of spectrum bands are in use for 2G. Overall, however, by far the most popular technology is GSM although it can be used in more than one spectrum band – often, in practice, 800 MHz rather than 900 MHz. The use of GSM is fostered by the factor of 'roaming' – that is, the ability of a GSM handset owner to take it to another country where it can still be used both internally and to make international calls to other countries where GSM is in use. Given the use of GSM throughout Western Europe, as well as its partial use in the USA, there is accordingly an obvious advantage for countries switching from analogue to digital to adopt it. However, some countries adopted one of the alternative standards to GSM before it became so well established. In particular, the US government made no attempt to enforce a national standard, and hence three different technologies were adopted. In addition to GSM, there was widespread use of both code division multiple

Table 1.1 Main licence holders – Western Europe[1]

Countries	Fixed wire	Mobile communications								
		Analogue	GSM[2]			PCNs[2]			UMTS	
Austria	Telekom Austria[3]	Mobilkom	Mobilkom[20]	7/92	2 800 000[e]	ONE[46]	8/98	1 300 000[g]	Hutchison 3G	11/00
			Max.mobil[21]	10/96	2 100 000[e]	Mobilkom	(under review)		Mannesmann 3G	
						Tele.ring [47]	5/00	100 000[f]	Max.mobil	
									ONE	
									3G Mobile[64]	
Belgium	Belgacom[4]	n/e	Proximus[22]	1/94	4 100 000[i]	KPN Orange[48]	6/99	1 004 000[i]	KPN Mobile 3G[65]	3/01
	Mobistar		Mobistar[23]	8/96	2 547 000[i]	Mobistar	11/00	n/e	Mobistar	
									Proximus	
Denmark	TDC[5]	n/e	TDC	3/92	1 680 000[e]	TDC	11/00		Hi3G Denmark[66]	9/01
			Sonofon[24]	3/92	998 000[g]	Sonofon	11/00		Orange	
			Orange	1/01		Orange[49]	3/98	520 000[e]	TDC	
			Telia	1/01		Telia	1/98	300 000[e]	Telia	
Finland	Sonera	Sonera	Sonera	6/92	2 361 000[g]	Sonera			Radiolinja	3/99
	Finnet Group[6]		Radiolinja	1/92	1 250 000[e]	Radiolinja[50]			Sonera	
			Suomen 2G[25]	1/00	n/e	Telia Finland	3/98	149 000[e]	Suomen 3G[67]	
						Suomen 2G			Telia Finland	
France	France Télécom	France Télécom	Orange	7/92	17 820 000[i]	Orange[51]			Bouygues Télécom	5/02
	Cégétel[7]		SFR[26]	7/92	12 560 000[i]	SFR			Orange	5/01
			Bouygues Tél[27]	5/96	6 620 000[i]	Bouygues Télécom			SFR	5/01
Germany	Deutsche Telekom	T-Mobile	T-Mobile[28]	7/92	23 100 000[i]	E-Plus[52]	5/94	7 480 000[i]	E-Plus Hutchison	8/00
	Mannesmann Arcor		Vodafone[29]	6/92	21 824 000[i]	Viag Interkom[53]	10/98	3 394 000[h]	Group 3G[68]	
	Viag Interkom					Vodafone			Mannesmann	
									MobilCom Multimedia[69]	
									T-Mobile	
									Viag Interkom	

Table 1.1 Continued

Countries	Fixed wire	Mobile communications								
		Analogue	GSM[2]			PCNs[2]			UMTS	
Greece	OTE	n/e	Vodafone[30]	7/93	2 884 000[i]	CosmOTE[54]	4/98	2 943 000[i]	CosmOTE	7/01
			Stet Hellas[30]	6/93	2 135 000[i]	Infoquest	7/01	n/e	Vodafone	
						Vodafone	7/01		Stet Hellas	
						Stet Hellas	7/01			
Ireland	Eircom[8]	Eircell	Eircell[31]	6/93	1 701 000[i]	Eircell	9/99			
			Digifone[32]	3/97	1 105 000[h]	Digifone	9/99			
			Meteor[33]	2/01	n/e	Meteor				
Italy	Telecom Italia	TIM[17]	TIM	4/95	22 650 000[g]	TIM			H3G (Andala)[70]	10/00
	Infostrada[9]		Omnitel[34]	12/95	17 431 000[i]	Omnitel			IPSE 2000[71]	
			Wind[35]	3/99	4 350 000[e]	Blu[55]	5/00	1 550 000[h]	Omnitel	
									TIM	
									Wind	
Luxembourg	EPT[10]	n/e	EPT	7/93	180 000[e]	EPT	3/99		EPT	5/02
			Millicom	5/98	123 000[e]	Millicom	3/99		Orange	
									Tele2	
Netherlands	KPN Telecom	n/e	KPN Mobile[36]	7/94	5 224 000[i]	KPN Mobile			Dutchtone	7/99
	Telfort[11]		Libertel[37]	9/95	3 231 000[i]	Libertel			KPN Mobile	
	Energis[12]		Dutchtone	1/99	1 000 000[e]	Dutchtone[56]			Libertel-Vodafone	
			Telfort	11/98	1 278 000[h]	Telfort			Telfort	
						Ben Nederland[57]	2/99	850 000[e]	3G-Blue[72]	

Norway	Telenor	Telenor[18]	Telenor	5/93	2 233 000[g]	Telenor			*Broadband Mobile*[73]	
			NetCom GSM[38]	9/93	850 000[e]	NetCom GSM			NetCom GSM	11/00
						Telia Norge[58]	10/00		Tele2 Norge[74]	
									Telenor	
Portugal	Portugal Telecom	TMN	TMN[39]	10/92	3 600 000[h]	TMN			ONI-Way[75]	12/00
	ETG[13]		Vodafone[40]	10/92	2 786 000[i]	Vodafone			Optimus	
			Optimus[41]	9/98	1 150 000[d]	Optimus			Telecel	
									TMN	
Spain	Telefónica	Telefónica	Telefónica	7/95	16 800 000[i]	Telefónica			Airtel	3/00
	Retevisión[14]		Vodafone[42]	10/95	7 637 000[i]	Vodafone			Amena	
	Uni2[15]					Amena[59]	1/99	5 220 000[i]	Telefónica	
									Xfera[76]	
Sweden	Telia	Telia Mobitel	Telia Mobitel	11/92	3 300 000[e]	Telia Mobitel			Europolitan	12/00
	Tele2[16]		Tele2[43]	1/92	2 010 000[e]	Tele2			Hi3G[77]	
			Europolitan[44]	9/92	1 100 000[i]	Europolitan			Orange Sverige[78]	
						Tele8 Kontakt[60]	n/a	n/a	Tele2	
Switzerland	Swisscom	n/e	Swisscom	3/93	3 510 000[h]	Swisscom			Dspeed[79]	11/00
			TDC Schweiz[45]	12/98	675 000[e]	TDC Schweiz			Orange	
						Orange[61]	6/99	800 000[f]	Swissccom	
									Team 3G[80]	
United Kingdom	BT	mmO_2[19]	mmO_2	7/94	10 968 000[h]	One-2-One[62]	9/93	10 400 000[i]	BT3G	4/00
	C&W	Vodafone	Vodafone	12/91	12 800 000[h]	Orange[63]	4/94	12 400 000[i]	Hutchison 3G[81]	
									One-2-One	
									Orange	
									Vodafone	
Isle of Man	BT		Manx Telecom		39 000[h]				BT	

Notes

1 Details obtained from *The Economist*, *Financial Times*, www.totaltele.com, www.cit-online.com and media cuttings.

2 The entries consist of name of operator, date when its service was first launched, and the number of subscribers where entries marked (e) relate to the end of 2000, marked (f) to the end of March 2001, marked (g) to the end of June 2001, marked (h) to the end of September 2001 and marked (i) to the end of December 2001. Where an operator provides both GSM and PCNs, the subscriber data are provided for both services together in the GSM column. There is some controversy over the counting of 'inactive' customers. For example, Vodafone and Orange in the UK delete customers who have been inactive (making, say, no outgoing calls and receiving fewer than four incoming calls per month) for three months, whereas some telcos only do so after a year of inactivity and some not at all.

3 Telecom Italia holds a 29.8% stake.

4 Majority owned by SBC (49.99%), TDC (16.5%) and Singapore Telecommunications.

5 Formerly known as Tele Danmark. SBC has a 41.6% stake.

6 Association of Local Telephone Companies.

7 Cégétel is owned by Vivendi Universal (44%), BT (26%), SBC (15%) and Vodafone (15%).

8 Acquired by e-Island, after divestment of its mobile subsidiary Eircell (see note 31).

9 Owned by Enel and merged with Infostrada giving France Télécom a 26.5% stake in the combined company.

10 Enterprises des Postes et Télécommunications.

11 Wholly owned by BT.

12 Formerly Enertel, now owned by Energis of the UK.

13 Commencing in 2000, and consisting, among others, of the national electricity, gas and railway companies.

14 A subsidiary of Auna which is owned by Grupo Endesa (29.9%), Unión Fenosa (18.7%), Banco Santander Central Hispano (23.5%) and ING (10.4%).

15 Formerly Lince, and owned by France Télécom (69%), Multitel (23.41%), Ferrovial (7.59%) and others.

16 Tele2 was previously NetCom AB (not to be confused with NetCom GSM in Norway, owned by Telia). A 12.73% stake is held by Millicom International Cellular which, like Tele2, is a subsidiary of NetCom AB. Tele2 acts as a mobile virtual network operator (MVNO) via the Sonofon network in Denmark, and via the Telfort network in the Netherlands.

17 Telecom Italia Mobile. Olivetti has a controlling interest via the 55% stake held by Tecnost in Telecom Italia (due to fall as savings shares are repurchased) which in turn holds roughly 55% of the voting shares in TIM.

18 Trading as 'Tele-mobil A/S'.

19 Trading as O_2.

20 Mobilkom Austria was spun off from PT Austria in May 1996. Telecom Italia has a 25% + 1 share stake.

21 Wholly owned by Deutsche Telekom.

22 Owned by Belgacom (75%) and Vodafone (25%).

23 Owned by Orange (50.7%), Telindus (5.0%), other financial investors (22.9%) and the public (21.2%).

24 Owned by Telenor (53.5%) and BellSouth (46.5%).

25 Owned by a consortium of 36 members of the Finnet Group.

26 Société Française du Radiotéléphone is owned by Vivendi Universal (35.2%), BT (20.8%),Vodafone (31.9%) and SBC (12%).

27 Bouygues Télécom is owned by Bouygues (64.5%), E.ON (17.5%), JC Decaux (11.5%) and BNP Paribas (6.5%).

28 The holding company is called T-Mobile International. In February 2002, T-Mobil DeTeMobil Deutsche Telekom MobilNet was renamed T-Mobile Deutschland.

29 Previously trading as D2, and 99.2% owned by, and now trading as, Vodafone.
30 Vodafone (rebranded from Panafon in January 2002) is mainly owned by Vodafone (52.8%), the public (36.6%) and Intracom of Greece (8.7%). Stet Hellas is majority owned by Telecom Italia (58.1%) and trades as Tele Stet.
31 Owned by Vodafone.
32 Owned by BT.
33 Owned by Western Wireless (60%), RF Communications (30%) and The Walter Group (10%). The licence was originally awarded in June 1998 but was not confirmed by the Irish Supreme Court until May 2000.
34 Omnitel Vodafone is owned by Vodafone (76.9%) and Verizon Communications (23.1%).
35 Owned by Enel and France Télécom. The merger of Infostrada and Wind was ratified at the end of 2001, at which point the Orange stake fell to 26.5% with an option to increase this to 32.5% when Wind has an IPO.
36 Owned by KPN Telecom (85%) and DoCoMo (15%). The brand name of KPN Orange is to be retained until June 2002 following on from KPN's purchase of the stake held by Orange – which now controls Dutchtone – after which it will switch to 'Base'.
37 Libertel-Vodafone is owned by Vodafone (70%), ING Bank (7.5%), private investors and employees.
38 Wholly owned by Telia.
39 Telecomunicações Móveis Nacionais is owned by Portugal Telecom.
40 Formerly Telecel. Owned by Vodafone (50.9%) and private investors.
41 Owners include the Sonae conglomerate (45%), Orange (20%), Telecommunicações Celulares (25%) and others.
42 Formerly Airtel. Owned by Vodafone (91.6%), Acciona (5.4%) and Torreal (3%).
43 Formerly NetCom AB.
44 Owned by Vodafone (71.1%) and private investors.
45 Owned by TDC (78.6%), diAx Holdings (16.6%), Swiss Federal Railways (2.6%) and UBS (2.1%).
46 Formerly called Connect Austria. Owned by E.ON (50.1%), Telenor (17.45%), Orange (17.45%) and TDC (15.0%).
47 Owned by Western Wireless International.
48 Owned by KPN Mobile.
49 Formerly called Mobilix. Owned by Orange (53.6%), the national railways (Banestyrelsen – 14%) and financial investors (32.4%).
50 A subsidiary of Elisa Communications.
51 Majority owned by France Télécom.
52 Owned by KPN Mobile (77.49%) and KPN Telecom (22.51%).
53 Wholly owned by BT and trading as 'city*partner*'.
54 Largely owned by Hellenic Telecommunications Organisation (OTE) (59%), Telenor B-Invest (18% of which Telenor 13.2%) and private investors (17%).
55 Consisting of Benetton (via stakes in Autostrade Telecomunicazioni (32%) and Edizione Holding (9%)), BT (29%), Distacom (9%), BNL (7%), Italgas (7%) and Caltagirone (7%). Various shareholdings are up for sale.
56 Owned by Orange. The Orange brand name will be recovered in September 2002 (see note 36 above).
57 A joint venture between Deutsche Telekom (50% – 1 share), Belgacom (35.3%) and TDC (14.7%).
58 Owned by EniTel.
59 Formerly Retevisión Móvil, now owned by Auna.

[60] Owned by the Telenordia consortium (BT, TDC, Telenor – all 33%). The proposed merger between Telia and Telenor delayed network roll-out.
[61] France Télécom (85%) leads a consortium also consisting of Swissphone Engineering AG (5%) and Banque Cantonale Vaudoise (10%).
[62] Owned by Deutsche Telekom.
[63] Majority owned by France Télécom.
[64] Backed by Telefónica
[65] Owned by KPN (85%) and DoCoMo (15%).
[66] Owned by Hutchison Whampoa (60%) and Investor AB (40%).
[67] Owned by 42 members of the Finnet Group and Tele2 AB.
[68] Owned by Telefónica (57.2%) and Sonera (42.8%).
[69] Owned by France Télécom and MobilCom (France Télécom 28.5%).
[70] Owned by Hutchison Whampoa (88.2%), Cirtel (1.83%), Tiscali (0.3%), San Paolo-IMI (5.64%), HdP (1.13%), Gemina (0.56%) and Franco Bernabe (2.26%).
[71] Owned by Telefónica (45.6%), Sonera (12.6%), Atlanet, Banca di Roma and others.
[72] Owned by Deutsche Telekom (50% – 1 share), Belgacom (35.3%) and TDC (14.7%).
[73] Owned 50/50 by Sonera and Enitel of Norway. The licence was returned.
[74] Owned by Tele2 AB (formerly NetCom AB).
[75] Owned by Electricidade de Portugal subsidiary ONI (55%), Telenor (20%), Grapes Communications and Iberdrola.
[76] Owned by Vivendi, Sonera and Actividades de Construcción y Servicios.
[77] Owned by Hutchison Whampoa (60%) and Investor AB (40%).
[78] Owned by Orange (51%), Skanska (10%), Bredbandsbolaget (34%), Schibsted (2%) and NTL (3%).
[79] Majority owned by TDC.
[80] Owned by Telefónica.
[81] Controlled by Hutchison 3G UK Holdings which is owned by Hutchison Whampoa (65%), KPN (15%) and DoCoMo (20%).

n/e = non-existent; n/a = not available.

Source: Compiled by author.

access (CDMA) and time division multiple access (TDMA). Furthermore, the PCN band, which is known in the USA as personal communications services (PCS), operates at 1900 MHz, with the result that GSM handsets acquired in Western Europe cannot be used without modification in the USA. Where roaming needs to be achieved between, say, GSM and TDMA, it is initially necessary to use 'plastic' roaming whereby the user's subscriber identity module (SIM) card is transferred to a new handset that works on the correct air interface, but, as discussed below, chipsets are being developed that will eventually enable handsets to be used with a variety of air interfaces.

In Japan, the personal handyphone system (PHS) was introduced at an early stage to utilise a dense network of low-powered, and hence cheap, base stations when in mobile mode. PHS handsets can also be used via a connection to a fixed-wire network when in the home or office, and PHS is hence roughly equivalent to the digital enhanced cordless telecommunications (DECT) standard used in Western Europe. Given the limitations of the PHS standard – for example, PHS handsets cannot be used in vehicles moving at anything other than low speeds – there are still large numbers of PHS handsets in use but they are being phased out and replaced by the more popular 2G handsets, predominantly using CDMA technology.

Third-generation mobile telephony (3G) is essentially an extension of 2G capabilities, although, as discussed below, the developmental process has in practice required an intermediate stage, known commonly as 2.5G, to be introduced. 3G, incorporating both terrestrial and satellite components, is known in Western Europe as the universal mobile telecommunications system (UMTS). UMTS has been described by ETSI as 'personalised globally accessible high quality mobile communications services', but it is more commonly described as 'mobile interactive multimedia services'. The official European Commission version is that UMTS is an advance on GSM capable of combining the use of terrestrial and satellite components.[1] UMTS does not strictly represent a major technological breakthrough because GSM already provides many services that were not originally expected to evolve until a later stage of development. Furthermore, upgraded or 2.5G variants of GSM will coexist with UMTS for an extended period. However, UMTS will eventually result in the provision of more sophisticated services that require greater bandwidth – effectively multimedia-on-the-move.

3G utilises a so-called 'packet-switched' network. The most crucial difference between a circuit-switched network – historically used for 2G but in the process of being phased out as speedily as possible – which requires a physical circuit path to exist between sender and receiver for the duration of a call, and a packet-switched network which is open to all users all of the time, is arguably that the latter enables subscribers to have an 'always on' connection to the Internet, and hence one that permits charges to be levied per packet sent rather than according to the time the circuit is open. However, the most noticeable aspect is concerned with the speed with which data traffic – becoming increasingly more important relative to voice traffic – is potentially downloadable. The basic packet-switched 2G normally operates only at between 9.6 and 14.4 kilobits per second (Kbps), whereas 3G should eventually operate at a minimum of 384 Kbps.

At the heart of a modern mobile network are to be found a number of base stations which link call-originating handsets to such mobile and fixed-wire networks as are capable of receiving the transmitted signals. The signals are picked up and passed on by antennas attached to the base stations, each of which sits at the heart of a cell covering a specific geographic area. Switching centres connect up the base stations, keep track of calls and ensure, for example, that if the sender is moving around that the call is picked up by the nearest cell. Base stations connect up to mobile switches either via fibre-optic cable or, if there is line of sight, by a radio link. In turn, a mobile switch may connect up to a fixed-wire switch in order to pick up or deliver a call to a fixed handset.

The size of a cell is dependent upon a number of factors, and in particular the radio frequency in use since the higher the frequency, the smaller the cell. A further issue of significance is the volume of traffic to be handled since a base station has a finite capacity, and when it is reached an additional base station will need to be built and the size of cells reduced to compensate. Also, physical features such as buildings or natural aspects of the terrain may need to be circumvented by creating additional cells. In general, urban cells will be located no less than 0.2 km and no more than 0.5 km apart, while rural cells will be ten times as far apart.

Since the spectrum licensed to a specific network is strictly limited, it is important that a particular frequency is utilised efficiently. This is done by taking account of the fact that a signal weakens as it travels, so although it will be strong in the nearest cell to an antenna, it will be very weak two cells further away. Hence, more distant cells can use the same frequency without fear of interference. Increasing the number of cells accordingly increases the number of times a frequency can be used within a geographic area.

The original analogue technology was subject to a number of limitations, most particularly its inability to be compressed without loss of clarity and the fact that, whereas it was satisfactory for voice telephony, it was unsatisfactory for data transmission which had to be converted from a digital to an analogue format prior to transmission and then back again at the receiving end. Analogue transmission tends to suffer from crackle, lack of security and vulnerability to fraud, although these shortcomings are partly compensated by its virtue of not fading out.

Digital signals were initially approved at the end of the 1980s – for example by the FCC in the USA in 1988. So long as these continue to coexist with analogue signals – there is no great incentive for subscribers to switch to digital for voice telephony alone, and hence there are, for example, still a number of analogue networks operational in the USA – most handsets need to be dual mode to cope with both types of signal.

The common digital system in the European Union, GSM, which was licensed widely from the beginning of 1992, uses the TDMA standard which allows each call to run along a channel within the available bandwidth and assigns it a fraction of the time available within that channel. However, in the Americas, the predominant standard is CDMA which assigns a special electronic code to each signal and hence allows the entire frequency band to be occupied simultaneously. This is claimed to provide greater capacity, better sound quality, lower power consumption and a decreased potential for fraudulent use.

Mobile networks are gradually changing over from a circuit-switched to a packet-switched basis. Packet-switching divides data up into individual packets of a specified size and in a specified format before sending them along the network. The correct sequence for delivery, and the address for delivery, are determined via control information attached to the packets before being sent. As a packet-switched network is permanently available to all data from any source, these inevitably get jumbled up together and may take a variety of routes from any one source to any one destination. However, the control instructions should ensure that the data is reassembled into the correct order before delivery. Since this may be subject to delays, it is advisable to keep voice data – which are time-critical – separate from non-voice data.

Mobile versus fixed-wire

The growth rate of fixed-wire subscriptions has been surprisingly steady since 1950. The 100 million mark was reached worldwide in approximately 1960 and the 1000 million mark was reached roughly at the end of 2001 when, according to the International Telecommunication Union (ITU), the total stood at 1045 million. Overall, there are as yet few signs of saturation. Although the most developed markets no longer have much growth potential other than via the installation of second lines for teenagers or computer links – penetration rose to 75 per cent early in 2002 taking Western Europe as a whole – penetration remains extremely low in some very large markets such as China and India. However, a further 10-times growth over the next four decades to 10 000 million is unlikely because it has become clear that it is uneconomic to run fixed-wire networks into areas that are either sparsely populated or where incomes per capita are low. Furthermore, there is already evidence that, in countries where mobile telephony use is at a high level, there is a tendency for households to cancel their fixed-wire contracts, although this is limited by the need to maintain a fixed-wire link to permit Internet access via personal computers (PCs).

As a consequence, the development of telecommunications in both wealthy and poor countries has relied increasingly upon mobile telephony either as a complement to, or substitute for, fixed-wire connections. Table 1.2 illustrates the rapid escalation in the number of countries where the number of mobile handsets has overtaken that of fixed ones.

During 2000, the number of mobile subscriptions rose by 230 million to a total of roughly 730 million, continuing a trend of 50 per cent

Table 1.2 Countries[1] where mobile handsets exceed fixed-wire handsets

1998	1999	2000		2001[2]
Finland	Austria	Bahrain	Philippines	Africa[3]
	Côte d'Ivoire	Belgium	Rwanda	Australia
	Hong Kong	Botswana	Senegal	
	Israel	Chile	Seychelles	
	Italy	El Salvador	Singapore	
	Japan	Greece	Slovenia	
	Paraguay	Iceland	South Africa	
	Portugal	Ireland	Taiwan	
	South Korea	Luxembourg	Tanzania	
	Uganda	Mexico	UAR	
	Venezuela	Morocco	UK	
		Netherlands		

Notes
1 It is reported that this was true of Cambodia as far back as 1993, but that may say more about the paucity of fixed-wire connections than of mobile usage.
2 Data for 2001 are not yet generally available.
3 Taken as a whole.

Source: Reports in media.

annual growth since 1996. At the end of 2001 the total was probably just over 900 million. As retail subscriptions suffered in the downturn for the telecommunications industry during 2001, it is somewhat hazardous to make predictions for demand in the future – there is also the need to factor-in near saturation in Western Europe where only 4.6 million new customers bought handsets in December 2001 compared to 11.9 million in December 2000 and 12.5 million a year previously, and a situation fast approaching saturation in North America, Japan and Korea – but by the end of 2002 it is reasonable to assume total retail subscriptions of well over one billion, roughly equivalent to a 25 per cent worldwide penetration rate – something of a slowing down compared to the previous growth rate but hardly a halt. The two billion mark may be reached somewhere around the end of 2006. Those who take a less conservative view prefer to believe that total mobile subscriptions will have comfortably exceeded fixed-wire subscriptions even as early as 2003. It is nevertheless important to bear in mind that subscriptions are not the same thing as handset sales. As is true of all technological product markets, the demand for handsets becomes ever more dependent upon replacements as time goes on, and this is especially true for handsets because they have become something of a fashion accessory. Total worldwide shipments of mobile handsets, expected to reach up to 600 million, only amounted in practice to 380 million during 2001, with Western Europe showing an absolute reduction – an unprecedented event. How rapidly demand will pick up again, and the extent to which it will be influenced by the switch over first to general packet radio service (GPRS)/cdma2000 1x and subsequently to 3G is difficult to say, but demand definitely will pick up again if only because of the constant introduction of new 'must have' handsets which tend to be introduced in cycles, although the net number of subscriptions in developed countries may fall close to zero by 2005. By implication, there will be a geographic realignment of demand, with Western Europe and the USA, which currently account for roughly one half of total sales, seeing their combined market share fall to around one third. Meanwhile, growth will predominantly occur in parts of Asia, and especially in China and India where, in the latter case, subscriber numbers at the end of 2001 stood at 5.5 million, up 75 per cent compared to the end of 2000. This, in turn, will cause the relative balance in sales to move away somewhat from the currently dominant GSM technology – GSM subscriptions reached in excess of 600 million at the end of 2001 and are forecast to reach 750 million by the end of 2002 – towards CDMA.

China is currently the largest mobile market in the world with 145 million subscribers at the end of 2001, compared to 180 million fixed-wire connections. During 2001, the number of subscriptions rose by roughly five million a month, overtaking the USA as the largest market in July at which point both had 120 million subscribers – the USA ended 2001 with 129 million. It is predicted that both fixed-wire and mobile subscribers in China will reach roughly 200 million at the end of 2002. Given that mobile penetration in China remains low at under 10 per cent, considerable growth potential exists in principle although average incomes among untapped sectors of the population are too low to suggest that operators will pursue them with much enthusiasm for too much longer. At present, there are only two such operators, the China Mobile Communications Corp. and the much smaller China United Telecommunications Corp. – both of which have listed arms in Hong Kong trading as China Mobile (HK) and China Unicom respectively. It is widely expected that the dominant fixed-wire operator China Telecom will be split into two halves – the southern continuing to trade as China Telecom while the northern, after merging with China Netcom and Jitong Communications, evolves into China Netcom Jitong – with both being awarded mobile licences.

However, China's importance lies not only in the size of its user market but in that of its handset manufacturing capacity. It is hardly surprising that Western manufacturers are moving production to a much cheaper country with a huge internal market, but recent predictions suggest that by 2005 as many as one in three handsets will be made there – perhaps 200 million compared to only 54 million in 2000 which represented 12 per cent of world output at the time. The increased output has its origins in joint ventures set up by Ericsson, Nokia, Motorola and Siemens in particular, but local manufacturers, supported by the government, claimed 15 per cent of the domestic market for themselves in 2001. What worries analysts is that the expected consequence of huge increases in output accompanied by intense competition and falling wholesale prices has recently manifested itself in the television set market, and profit margins are expected to become dangerously low even as China takes over the lion's share of worldwide production.

A further issue to be noted is that although a specified mobile network operator's status within a national market is of considerable importance, there are clearly economies of scale to be derived via worldwide coverage, and, further, that owning networks in a variety of countries is a clear source of such economies because it eliminates the need

Table 1.3 Mobile operators' controlled subscribers

Operator	Subscribers – 31/12/00	Subscribers – 31/12/01
Vodafone Group	78 700 000	99 900 000
China Mobile (Hong Kong)	45 100 000	69 200 000
NTT DoCoMo	35 950 000	48 600 000[1]
China Mobile	31 830 000	45 000 000[1]
T-Mobile	31 000 000	51 200 000
Orange	30 460 000	39 300 000
Verizon Wireless	27 500 000	29 200 000
Telecom Italia Mobile	27 400 000	35 000 000[1]
Telefónica Móviles	23 200 000	29 800 000
mmO_2	20 700 000	17 240 000
Cingular Wireless	19 680 000	18 000 000
AT&T Wireless	15 710 000	20 800 000[1]
KDDI	13 710 000	27 030 000
China Unicom	12 800 000	27 030 000
KPN Mobile	12 100 000	13 710 000
SK Telecom	10 930 000	15 500 000[1]
Sprint PCS	10 660 000	14 000 000
América Móvil	10 300 000	26 000 000[1]

Note

[1] Estimated via extrapolation from last declared figure.

Source: Company announcements.

for roaming arrangements. Hence, it is of interest to examine Table 1.3 which sets out the number of subscribers controlled by individual operators, where the number of subscribers controlled in each market is a pro rata of total subscribers based upon the percentage of voting shares or overall equity as appropriate. It is worth observing that the extent to which the operators are truly worldwide is hugely variable. Whereas such an assertion is applicable in the case of Vodafone where only roughly 13 million subscribers are domestic – that is, 13 per cent of the total – it does not apply at all in the case of the Chinese operators which are 100 per cent domestic and intend to stay that way. Nor is it possible to generalise overmuch by region. While it is true that Asian operators are locally focused almost without exception, US operators may be either fairly equally balanced – as in the case of Cingular Wireless – or very domestically biased – as in the case of AT&T Wireless. The contrast is most interesting in Western Europe where the likes of T-Mobile have been expanding heavily overseas while at the same time the likes of mmO_2 are pulling in their horns, although current levels of indebtedness suggest that expensive overseas forays will be a thing of the past for some time. It may also be observed that there is as yet no operator listed from India which would appear to have enormous potential in terms of population size. Finally, it may be noted that where operators' controlled subscribers have grown sharply it may either be accounted for by domestic growth, as in China, or by takeovers as everywhere else – for example, Deutsche Telecom took over VoiceStream in the USA during 2001, thereby gaining over seven million subscribers – and that sharp reductions are accounted for by disposals, as in the case of mmO_2 (the hived-off wireless subsidiary of BT), recently rebranded O_2.

Short message service (SMS)

So far the emphasis has been upon voice telephony, which is understandable since that is what handsets were originally created to sustain. Nevertheless, it is worth noting, since the issue also has to be addressed in the context of 3G services, that increasingly handsets are being used to convey short messages.

For those brought up on full 'qwerty' computer keyboards, the idea of tapping out messages on a tiny mobile handset keyboard, where each number also represents several letters and symbols, is little short of purgatory. However, for those so disposed – predominantly teenagers who are given mobile phones at an early age, ostensibly so that they can keep their parents informed of their whereabouts – it is only a matter of practice before messages can be tapped out at great speed – it can even be done under the desk in the classroom. Furthermore, given limitations upon the length of a message that can be sent for a standard rate – usually 160 characters – it has become the custom to use short symbolic abbreviations or representations for quite complex ideas – in effect designing a new form of language.

SMS – sometimes known as short messaging – is not particularly cheap, and does not accordingly seem worthwhile to those with monthly contracts who can often talk free in off-peak periods. However, most teenagers have prepaid handsets where SMS represents an economic means of communication compared to non-focused conversation – which may anyway not be practical in noisy bars and nightclubs. In addition, businessmen can send urgent messages to colleagues in meetings where decorum requires that handsets are put in silent mode. SMS is increasingly being used to, for example, send messages on special occasions – Valentine's Day messages can be sent without disclosing the number of the sender. As a result, it has taken off successively in various parts of the world. The worldwide totals at the end of 1999, 2000 and 2001 respectively were roughly 4, 20 and 30 billion a month. SMS initially became very popular in Europe – one half of all Scandinavians use it and T-Mobile alone carries nearly one billion messages a month in Germany – and in parts of Asia. Meanwhile, it barely signified in the USA where 'instant messaging' between computers was commonplace and local calls generally free. Despite this, SMS has recently become a phenomenon in the USA with some 400 million messages sent in December 2001, and it is rapidly making inroads into both instant messaging between computers and sending e-mails via the wireless application protocol (WAP). This is partly accounted for by the introduction of SMS interoperability between networks, initially introduced by AT&T Wireless in November 2001 but yet to become all pervasive – Verizon Wireless has yet to join in. When all operators are party to interoperability the US SMS market is expected to become the largest in the world given that AT&T Wireless is already seeing 30 per cent of its subscribers' messages being sent to other networks' subscribers. Another factor sustaining SMS growth in the USA is the steady upgrading of handsets to those that are capable of two-way messaging.

SMS has been a real boon to operators because it is very cheap to set up. Unlike voice, messages can be stored in short message centres and delivered when the networks have spare capacity. Operators earn revenue at the rate of $0.10 or more per message – in other words, they are currently earning something like $40 billion a month from this service overall and possibly as much as 10 per cent of an individual operator's revenue is derived from SMS with a profit margin of at least twice that for voice telephony. Furthermore, text messages can be used for premium-rate services – the downloading of ring tones is especially popular. SMS is due to evolve into a multimedia messaging service (MMS) capable of conveying audio and video clips, photographs and images (known as m-postcards) as the capabilities of the networks are upgraded through 2.5G to 3G. It should be borne in mind here that a still image only requires 50 KB of memory whereas a 30-second video requires 3 MB. It is estimated that a multimedia message will cost between $0.50 and $1.50. Handsets with built-in digital cameras became available in Japan during 2001, and proved to be a major factor in J-Phone's increase in market share during the latter part of the year. In Europe, a form of MMS called 'SMS TV Chat' was introduced in Germany by T-Mobile in July 2001 and generated nine million messages in its first six months of operation. However, European handset manufacturers have been somewhat slow off the mark in getting MMS-capable handsets onto the market. Market leader Nokia, for example, will only have them available towards the middle of 2002 at which point the likes of Sonera intend to launch a MMS service, although Nokia expects the market to reach critical mass before the year-end – a quite different experience from the relatively slow build-up of SMS.

National coverage: Western Europe versus the USA

Western Europe and the USA do not differ exclusively in terms of the technological developments outlined above. A further important factor in relation to the move towards 3G is the issue of national coverage. In Western Europe, all mobile licences are national, and both geographic and population coverage are expected to be very extensive, especially in the 900 MHz band. Obviously, individual operators do not tend to have coverage in every country, but there is a clearly-established system for generating roaming agreements that is much simplified by the existence of common technology. Roaming may not be cheap, for reasons discussed below, but it is commonplace.

National networks are not, however, the norm in the USA. Initially, the customary practice was for the FCC to issue two licences in each designated area, one to the incumbent Regional Bell Operating Company (RBOC) and one to a competitor via a tender. In such a world, licence ownership was necessarily fragmented and roaming arrangements rather complicated. Indeed, it took approximately a decade for automated roaming to be implemented. In the mid-1990s, this system was subjected to significant pressure for change. In part this stemmed from the 1996 Telecommunications Act, but, in addition, the commencement of a series of auctions for PCS spectrum provided the RBOCs with an opportunity to expand outside their initial boundaries. Importantly, they were not the only ones to benefit since the long-distance operators could also now attempt to build extensive mobile networks.

One of the predictable consequences was for a number of operators to attempt to buy their way to something like national coverage. However, while this could theoretically be done directly via licence acquisition, it often appeared to be more rational to acquire companies that already possessed licences because suitable licences were not always available in the auctions or were too costly because of competitive bidding in metropolitan areas. AT&T, for example, accordingly pursued both strategies, combining the purchase of the outstanding 80 per cent of McCaw Cellular in 1994 to form AT&T Wireless with that of 'A' and 'B' block metropolitan licences in 1995 and 'D', 'E' and 'F' block local licences in 1997, at which point it could in principle reach a population of 250 million.

The RBOCs initially set out to ensure full coverage of their domestic regions, but subsequently expanded via mergers, acquisitions and joint ventures both among themselves (e.g. SBC/Pacific Telesis forming SBC Communications), with non-RBOCs (e.g. Bell Atlantic/GTE forming Verizon Communications), with foreign operators (e.g. Verizon Communications wireless/Vodafone forming Verizon Wireless) or with smaller local operators (e.g. SBC/Comcast Cellular). Meanwhile, existing mobile operators concentrated upon acquiring more spectrum in the auctions – hence leading to the apparent debacle in the 'C' licence auction when NextWave and Omnipoint were unable to pay for their licences. Omnipoint was subsequently swallowed up by VoiceStream Wireless, itself spun off Western Wireless, another bidder in the auction, only for VoiceStream to itself become a subsidiary of Deutsche Telekom.

Some attempt has been made to address the roaming issue via a distinction between 'on-net' and 'off-net' roaming. In the former case, a Verizon Wireless customer in one area can, for example, interconnect with a Verizon customer in another area without the payment of an interconnection fee. However, smaller operators may need to pay interconnection fees to larger operators even in their home regions if they do not have complete coverage. The ability to keep prices down by eliminating interconnection fees is obviously a major incentive for operators to achieve national coverage.

However, there is an additional factor to take into account. Within Europe, mobile telephony has developed as a form of premium-priced product subject to much less regulatory control than incumbents' (initially monopolised) long-distance and local fixed-wire networks. In comparison, the distinction is not regarded as very relevant in the USA where local calls are generally provided free and, in any event, people often keep their mobile handsets shut off to avoid being charged for incoming calls. As a consequence, there is nothing like the profit margin in mobile telephony in the USA and, if high roaming charges are levied, these cannot readily be passed on to subscribers, thereby eroding even these margins. At the same time, local operators have no incentive to reduce interconnection charges to larger operators seeking a national footprint since they are not in a position to offer national roaming themselves.

One possible way for such smaller operators to at least give the impression that a national footprint exists is to provide services under the umbrella of a common brand (e.g. Cellular One). However, even this is problematic to organise unless, as in the case of Cellular One, the individual companies are all subsidiaries of a handful of companies (in this case AT&T Wireless, Cingular Wireless and Western Wireless). The biggest operators also tend to link up with affiliated companies in the PCS market, often using them as bidders in auctions where preference is given to new entrants. Some of these may also be swallowed up over time (e.g. in the case of TeleCorp by AT&T Wireless in October 2001). It may be noted as a final point that if local operators' names appear to have value they may be preserved even if a more or less national tariff structure exists under a common name not linked to any of the individual companies.

Hence, in summary, the issue as to whether an operator provides national coverage is difficult to pin down. Certainly, the kind of national coverage encountered in Europe is not currently being provided, although it is fair to say that no individual European market is anything like the size of the USA and pan-European coverage cannot be provided by any single operator without roaming agreements. However, there appear to be six potential candidates, namely AT&T Wireless, Cingular Wireless, Nextel Communications, Sprint PCS, Verizon Wireless and VoiceStream Wireless. It may be noted that the latter is now European owned and that Verizon Wireless is minority owned by Vodafone.

ARPUs and SACs

Increasingly, as mobile markets reach saturation, attention is being drawn to the issues of average revenue per user (ARPU) and subscriber acquisition costs (SACs). It is a straightforward, if regrettable, fact of life for network operators that as penetration rates increase there is an associated tendency for SACs to rise and ARPUs to fall. In part this is because of ever-increasing competitive pressure and in part because there is a tendency for new subscribers increasingly to sign on for prepaid rather than annual contracts. This tendency, in turn, has been driven in many countries by the fact that a prepaid subscriber has merely to acquire a handset and a prepaid card to enjoy the benefits of mobile communications, and the amount of expenditure can be strictly measured. It is therefore highly suited to those who have temporary addresses and no credit history – in other words the likes of students who represent an increasingly large part of the market. Although unsubsidised handsets may be expensive, it is often the parents who buy them so that their children can keep in touch.

Until fairly recently the financial markets were interested only in the growth rate of subscriber numbers, assuming that high growth rates would deliver higher revenues and profits. It is now evident that beyond a certain point – there is evidence that this occurs when penetration reaches roughly 20 per cent – the ARPU tends to decline quite rapidly, and that if an operator wants to maintain the numbers on annual contracts, it must provide subsidised handsets. The latter policy is not necessarily uneconomic provided call charges can be maintained, but if either the pressures of competition or regulatory interference drives them down the situation can deteriorate rapidly.

The proportion of prepaid subscribers is hugely variable between countries for a variety of reasons, including whether prepaid was introduced before annual contracts had become the norm. In Italy and Portugal, for example, all operators have prepaid proportions of 70 per cent or more, whereas in the UK it is nearer 50 per cent individually but over 60 per cent overall and in France below 50 per cent overall. In contrast, only about 10 per cent of US subscribers use prepaid and the figure for Japan is below 5 per cent.

The tendency for ARPU to decline at a rate of 5–10 per cent a year is related to the tendency for voice telephony use to stop growing. In the short term, as noted elsewhere, ARPU is being bolstered by SMS and the initial trickle of data transport, but the latter is going to have to increase significantly even to maintain current levels of ARPU. As discussed in the section below on i-mode in Japan, its introduction has raised ARPU by roughly $10 to a little above $60 a month, and it is obviously hoped that the introduction of GPRS will achieve something similar, although that did not happen with WAP. However, even $50 is more than can currently be achieved in most countries other than Japan – $40 a month is a more typical figure in the USA and rather less in Western Europe.

Furthermore, SACs for 2G are one thing, for 3G quite another because 3G requires a great deal of up-front investment that must ultimately be compensated by higher revenues. The size of the problem is very variable by country, if only because, as detailed in Chapters 3 and 4, the licensing costs and roll-out conditions have been determined on an individual country basis even within the EU. Where licences are free and roll-out relatively cheap, as in Japan, the prognosis is generally good. Where licences are expensive, and roll-out costs relatively high, as in Germany, the prognosis is much worse unless, of course, some of the licensees fall by the wayside since the number of licences is obviously going to be a further significant aspect of the ability of licensees to make profits from 3G.

Evolution of standards

The issue of technical standards clearly needs to be resolved if global interconnection is to be achieved. In late 1997, the ITU published a report – *Framework for Modularity and Radio Commonality within IMT-2000* – which advocated the introduction of a standardised system within the context of International Mobile Telecommunications-2000 (IMT-2000) by the year 2002.[2] The idea was that both carriers and equipment manufacturers would be provided with a set of common building blocks with which any third-generation system had to be made compatible. This was a critical aspect since it signified that there was not to be a unique standard but rather a 'family of standards' which would allow all existing networks, with their attendant historic heavy investment, to be modified rather than discarded. The actual equipment could take a variety of forms, including mobile handsets with a tiny TV screen or a telecommunications link to a palmtop or laptop computer.

Technically, a 3G network could utilise almost all spectrum lying between 400 MHz and 3 GHz. Despite this potentially vast spectrum range, it had already been determined, as a result of a decision made in 1992 by the World Radiocommunications Conference (WRC) that the common frequencies in all third-generation systems would be restricted to 1885–2025 and 2110–2200 MHz (see Figure 1.1). This would make international roaming between networks a reality, together with the transmission of data at potentially up to 144 Kbps on the move and 2 Mbps (megabits or million bits per second) while stationary. This was particularly important because whereas the transmission of mobile data

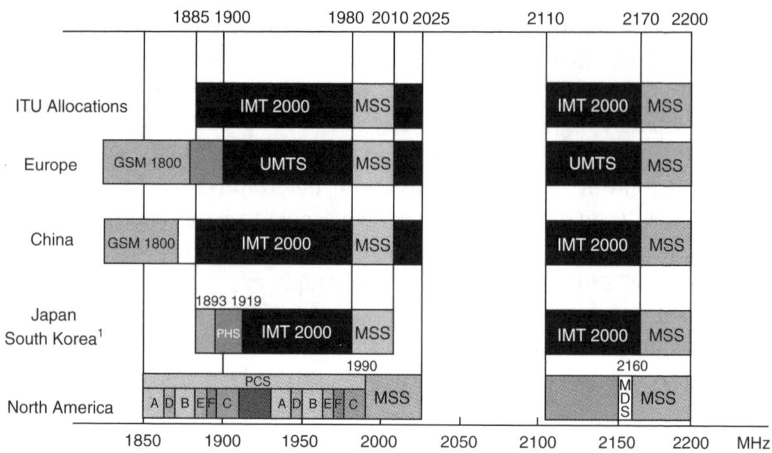

Figure 1.1 3G spectrum

Note

[1] There is no PHS in Korea.

MSS = Mobile Satellite Services; MDS = Mobile Data Service/Multipoint Service.

was not expected to be relevant to more than 25 per cent of wireless subscribers even by 2010, it was likely to represent at least 50 per cent of carriers' profits. However, the decision on the preferred frequency ranges did lead some commentators to wonder whether the technical advantages derived from a move to new spectrum for mobile communications were greater than the drawbacks arising from the need to design, build and test new networks from scratch, build additional base stations and deal with the need to migrate progressively from 2G networks using lower frequencies.

In June 1997, the European Radiocommunications Committee (ERC) decided (ERC/DEC/(97)07) to divide up the WRC frequency bands, allocating 1900–1980, 2010–2025 and 2110–2170 MHz to the terrestrial universal mobile telecommunications system (UMTS) as well as 1980–2010 and 2170–2200 MHz to satellite applications which use a space division multiple access (SDMA) interface, but recognised that this would probably be insufficient and that additional spectrum would almost certainly need to be freed or reallocated by 2005 within the 900, 1800 and 1900 MHz bands used most commonly by 2G systems.[3]

In a further move, the ITU divided the new system into two parts based respectively upon the core network architecture and the radio (air) interface – linking base station and handset and known as the universal terrestrial radio access (UTRA) – and specified common components in each case. The telecoms industry was then left to work out how best to meet the requirements of each. So far as the core architecture was concerned, the choice was fairly limited in practice given the predominance of GSM in Europe, and cdmaOne (IS-95) and TDMA (IS-136) in the USA – the two respective core architectures are generally referred to as GSM-MAP and ANSI-41. The ITU wanted everything to be settled by the end of 1999 so that services could commence in 2002.[4]

Nokia, Ericsson, Alcatel and Siemens, Europe's leading mobile handset manufacturers, announced that they would be backing GSM-MAP for the core architecture. For the radio interface the choice lay essentially between wideband CDMA (W-CDMA) and a hybrid wideband TDMA/CDMA (W-TD/CDMA). Whereas Japan opted for W-CDMA throughout – effectively the same as for GSM-MAP in the core architecture – the UMTS approach to the radio interface contained elements of W-TD/CDMA. Work was set in hand to bring these into close compatibility.

The issue of compatibility with standards in the USA presented the most awkward problems because Lucent Technologies and Qualcomm wanted to use a somewhat different technology known as cdma2000 (sometimes known in its 3G guise as 3xRTT – Radio Transmission Technology). The USA was determined not to miss the boat as it had in the case of GSM, mainly because it had chosen not to enforce a single common standard but to allow market forces to resolve the issue, and wrote to the European Commission to warn it against taking pre-emptive action. The Commission replied that it would let the industry have the last word on the matter. It should be borne in mind in this connection that it was only during the third quarter of 1998 that sales of digital handsets exceeded sales of analogue handsets in the USA.

The present position is that there are three main radio interface modes within IMT-2000, known respectively as IMT DS (Direct Sequence or Spread), IMT MC (Multi-Carrier) and IMT TC (Time Code), which between them will be compatible with GSM, CDMA and TDMA. These modes are known more commonly as direct sequence frequency division duplex (DS-FDD – a type of CDMA (W-CDMA) supported by GSM carriers, which uses paired spectrum and is sometimes referred to as UTRA FDD); Multi-Carrier Frequency Division Duplex (MC-FDD – which uses paired spectrum and is based on cdma2000); and Time Division Duplex (TDD – which comes in a version known as UTRA TDD or one harmonised with the new TD-SCDMA standard proposed by China, both of which use unpaired spectrum).[5] In addition to these, the ITU has established a specification for IMT SC (Single Carrier), also known as UWC-136 (which, like W-CDMA, is a FDD system related to EDGE – see below), and for IMT FT (Frequency Time), commonly known as DECT. The forecast speed of the new technology is claimed to be 384 Kbps in urban areas and up to 2 Mbps in indoor locations, although these speeds are viewed with a certain amount of scepticism.

At the World Radiocommunications Conference 2000,[6] governments of 150 countries again addressed the issue of spectrum allocation in the hope of agreeing to a single global range for further expansion after 2005, essentially consisting of 160 MHz in one continuous band in the 2.5 GHz (2500 MHz) band. However, the best that could be agreed was a choice of three frequency bands, namely 806–960, 1710–1885 and 2500–2690 MHz – the latter heavily favoured by European organisations because it is the only one not currently occupied for 2G. For its part, the USA refused to commit itself to any of them because much of all this bandwidth was already occupied. This meant that handsets would need to have a built-in capacity to roam between frequencies and would probably be heavier and costlier than was desirable.[7] To ameliorate matters somewhat, the WRC also accepted that 3G networks could be built using existing 2G spectrum.

It should be noted that the chosen frequency bands for 3G are much higher than those used for GSM and this means than a 3G network needs far more cells than a GSM network because the signal range decreases as the frequency rises. Furthermore, higher frequency signals travel in straighter lines, which makes little difference in urban areas but a good deal in open rolling countryside. As a rule of thumb, a 3G network requires twice as many cells as a GSM network, but that ratio may prove to be unnecessarily high in practice, especially if 3G facilities are built on top of 2G base stations.

The issue of standardisation of W-CDMA/UMTS is currently being addressed by the third generation partnership project (3GPP).[8] A major set of specifications was published in mid-December 2000, with amendments agreed in March 2001, and the fact that specifications are constantly being updated is somewhat problematic because of the need for networks developed at different times, in different countries and by different equipment makers, to permit international roaming. This, in turn, has some bearing on realistic deadlines for network roll-outs (see, for example, Japan below).

The deadline for member states of the European Union to introduce an authorisation system for UMTS was initially specified by the European Commission[9] as 1 January 2000, and for harmonised provision of UMTS services as 1 January 2002, although the European Council added an amendment in September 1998 to the effect that member states which have 'exceptional difficulties in adjusting their national frequency plans ... can obtain an extra period of up to 12 months' beyond these two deadlines.[10] In addition, by February 1999, member states were expected to devise plans for the provision of such additional spectrum as would be needed. Most countries will need to shut down their analogue services and, if necessary, remove the relevant spectrum from organisations such as the armed services that have traditionally used it. A final factor is that the frequencies allocated to UMTS may have been set aside by other parties for the provision of satellite or other services such as PCNs.

For the moment, no-one knows precisely when UMTS services will become commercially available on a reasonably widespread basis, nor, for that matter, whether there will be any real demand for them, especially among the general public, particularly given the generally high level of satisfaction with services provided over GSM networks and the fact that the scope of those services is constantly being upgraded. The current GSM speed of 9.6 Kbps can be raised in principle through the successive implementation of HSCSD (high-speed circuit switched data) capable of handling between 38.8 and 57.6 Kbps using four bonded GSM circuits; GPRS, capable of up to 168 Kbps using eight 21 Kbps GSM time-slots, which will allow subscribers to remain permanently online and be billed only for the amount of data received – effectively 2.5G; and EDGE (enhanced data [rates] for GSM evolution) nominally capable of 384 Kbps through the expedient of simultaneously feeding bundles of spare capacity to a single handset.[11] Although EDGE can potentially compete in speed with 3G – since the latter is a shared provision the speed of delivery slows as the amount of data conveyed grows – it may require additional towers to compensate for its high frequency range as well as new handsets, and may ultimately be ignored by most operators given that the time frame for its introduction is similar to that of UMTS.

These intermediate stages between 2G and 3G may be as far as some carriers wish to go, and hence it is probable that several variants of 2.5G will be available alongside UMTS. In May 1999, for example, Lucent announced the introduction of PacketGSM that allows operators to launch 3G-style services over a modified 2G network. Such developments were expected to discourage some potential bidders from seeking 3G licences. However, some operators were becoming desperate for additional spectrum because the number of subscribers to their existing 2G networks was growing very rapidly, and hence they had no option other than to seek 3G licences. In other cases it was anticipated that non-licensees could apply to licence holders to become mobile 'virtual' network operators (MVNOs).[12]

As hypertext mark-up language (HTML – the language used in computer-based networks to access the Internet) was regarded in Europe as too complex for mobile handsets given their slow operating speeds compared to PCs, a simplified language, the wireless mark-up language (WML), was developed to improve download times and presentation. Nokia, Ericsson and Motorola linked up with Unwired Planet (renamed Phone.com) of the USA to promote the wireless application protocol (WAP), a technology that enables stripped down Internet web pages to be viewed on a mobile phone with a somewhat enlarged screen. Nokia announced its 7110 media phone, the world's first WAP-standard product, in July 1999. A thumb-operated rollerball allowed users to scroll through text. The service was slow to download and slower still because each individual service needed to be connected to a different information source or Web server – there was no 'always on' facility. Furthermore, a subscriber could not roam around the World Wide Web, but was restricted to proprietary text-based services (the so-called 'walled garden') chosen by the operator. The state-of-the-art handset in early 2000, the Motorola P1088, permitted access to the normal Internet and the mobile Internet, the sending and receiving of e-mails and text messages, and the sending of faxes. However, graphics were rather limited.

It is important to note that in Japan, where computer use and Internet access via PCs is low compared to Europe, DoCoMo chose to go straight to the so-called 'i-mode' solution whereby a new packet-switched network, based upon the PDC-P standard, would permit the use of compact HTML (cHTML). As a result, despite the fact that it could not deliver the full range of services anticipated after the introduction of 3G networks, DoCoMo appeared to have stolen a march on its European rivals. However, it seems unlikely that most European carriers will want to introduce i-mode as such given their investments in WAP, GPRS and UMTS. The only reasonable probability is that i-mode services will be introduced by KPN Mobile, in which DoCoMo has a 20 per cent stake, sometime in the Spring of 2002. KPN Mobile announced in December 2001 that it had reached agreements with one hundred national and international content providers and that some sixty certified sites would be available at launch, although non-certified providers would also be accessible on independent sites. The NEC handsets will be GPRS compatible and the built-in i-mode browsers will be capable of reading HTML and WML content.

The longer-term ambition is to make everything work using a common syntax, the eXtensible HTML (xHTML). In the meantime, there is something of a struggle going on between software developers who favour cHTML because it is relatively easy to adapt from HTML, and carriers who see i-mode as existing in a technological cul-de-sac.[13]

Mark-up languages

The situation with respect to mark-up languages is somewhat confusing, and a short clarification is therefore in order. As noted above, the dominant mark-up languages are currently WML – mainly applied to WAP-enabled handsets – and cHTML – mainly applied to i-mode-enabled handsets. WML has come in a number of versions, of which the most recent was WML 2.0 in August 2001. HTML, the computer language which began to be developed in 1990, is increasingly seen as having outlived its usefulness, in part because it is focused on displaying documents rather than manipulating data, a view also taken of the hand-held device mark-up language (HDML), with future development increasingly tied to xHTML. xHTML, in version 1.0, combines all elements of HTML 4.01 with the syntax of WML – itself linked to the eXtensible mark-up Language (XML). The great virtue of XML compared to HTML is its flexibility. In XML, the content structure is separated from presentation so an XML document either describes content *or* sets its structure. When a style sheet is attached to XML it can be converted to HTML for display on the Web. The development of xHTML Basic, originally approved in

2000, which brings together WAP and WML to suit mobile devices with limited memory, underpinned the recent release of WAP 2.0 in an attempt to overcome the incompatibility problems with different versions of WML. By retaining much of the syntax of HTML, problems are minimised for content developers, and xHTML Basic provides a common way forward for both WAP and i-mode. However, the constant reworking of mark-up languages does raise awkward issues of backwards compatibility, since the need to encompass more than one mark-up language in a single handset will make heavy demands upon the available memory.

The role of satellites

Although this text is primarily concerned with terrestrial links, it is probable that satellite links will have some role to play in the provision of 3G services, albeit mainly outside the so-called metropolitan area networks (MANs). It is possible that some of the spectrum allocated either for global mobile personal communications by satellite (GMPCS) in the 2 GHz or below range, or to Ka-band operators in the 18–30 GHz range, which was subsequently authorised for IMT-2000 use, will become reusable because the satellite systems originally awarded the spectrum will not be in orbit within the required seven-year period. However, although the 2 GHz band is compatible with terrestrial links, the high-frequency Ka-band presents considerable problems. Furthermore, a satellite system has to acquire the same spectrum in all countries covered by its footprint if it is to operate successfully, and the vagaries of licensing proposals for terrestrial 3G suggest that all kinds of incompatibilities may arise where satellite communications are concerned.[14] The existing geostationary operator, Inmarsat, is well placed to lead the way in 3G provision, and has recently ordered three new satellites to support the launch of a so-called broadband global area network (B-GAN) in 2004, which will in principle be able to transmit at 2 Mbps, possibly to terminals the size of personal digital assistants (PDAs). Ericsson was appointed as sole supplier of the network in December 2001, with a provisional launch date of 2004. Prior to that time it will be necessary to upgrade older satellite systems, which can currently only provide a 9.6 Kbps link (or up to 64 Kbps if multiplexed), as well as those planning to launch in the near future to a theoretical maximum of 600 Kbps. One facility which satellite operators will be able to develop is a 3G service to transoceanic aircraft. The somewhat chequered history of satellite systems is set out in more detail in Chapter 2.

2
Intermediate Steps Along the Road

This chapter addresses issues raised in Chapter 1 in more detail. First, it looks at the history of satellite provision. As indicated earlier, this is no longer expected to play a major role in mobile communications despite the high hopes, not to mention heavy investment, of five years ago. It is a salutary story in that it emphasises the point yet again that the development of technology tends to outrun the delivery of working networks based on existing technology, so that what appears to be an entirely sensible, economic proposition one year can prove to be an economic disaster a mere couple of years later.

The following sections deal with what is known as 2.5G. In Europe, 2.5G started out as WAP and has evolved in conjunction with GPRS. GPRS is now with us, but has taken time to become established. Although almost all mobile networks will become GPRS-enabled by some point in 2003 – or enabled by its equivalent, cdma2000 1xRTT – not all will necessarily move on to 3G. Equally, whereas i-mode is well-established in Japan where 3G will definitely become the norm, its role elsewhere, either as an intermediary stage to 3G or as an end point in its own right, is far less clear.

The chapter concludes with a brief review of personal digital assistants (PDAs). So far, the mobile terminals in common usage are almost all some variant of a standard handset. However, such a device has limitations in a broadband world given, for example, the restricted screen size and the need to key-in inputs using the multi-purpose keys on a handset. These limitations can readily be removed if a full 'qwerty' keyboard can be utilised in some way or instructions can be given without using a keyboard at all, in conjunction with a full-length display – precisely the virtues of PDAs. On the other hand, these cannot get progressively smaller and lighter until the point is reached where they can be slipped unobtrusively into, say, a trouser pocket.

Satellite communications

Introduction

Satellite communications are more correctly known as the global mobile satellite systems market (GMSS) where satellite personal communications services (S-PCS) are provided.[15] Alternatively, the term global mobile personal communications via satellite (GMPCS) is in usage.[16] The definition, agreed at the GMPCS MOU meeting in October 1997, is:

> any satellite system (i.e., fixed or mobile, broadband or narrow-band, global or regional, geostationary or non-geostationary, existing or planned) providing telecommunications services directly to end users from a constellation of satellites.

Satellite communications, often using a very small aperture terminal (VSAT)[17] that picks up signals via a dish antenna, can carry voice, data and video. The service can be one-way, for example the transmission of financial data by Reuters, or two-way involving telephony for the return path. Satellite communications form a necessary part of a fully developed mobile telephony system because of the need to extend linkages beyond the range of economically viable land-based networks. However, some analysts see their main potential lying in their ability to provide high-speed data delivery and multimedia, including Internet provision, to company branches.

In a terrestrial system, either the call recipient is using a fixed link to the public telephony system or both sender and receiver have to be in range of an earth station, the size of which has shrunk as a result of modern technology but remains fairly bulky. In a satellite system, where the receiver is not within range of a base station, the signal is diverted through a 'gateway' and bounced down to the receiver via a satellite. In some satellite systems, the signal is passed between satellites until it reaches the gateway nearest the intended call recipient. This serves to reduce the number of gateways needed to ensure full coverage, but it is a matter of commercial judgement whether to opt for additional satellites or additional gateways.[18]

Historically, satellites were in a geostationary, and sometimes geosynchronous, orbit (hence GEOs) – that is, they appeared to be stationary when viewed from the surface of the earth – and only certain parts of the globe were covered. Their stationary appearance reflected the very large distance from the earth at which they needed to be located, which, in turn, meant that there would be a delay in signal transmission – rendering conversation a rather erratic affair. A GEO's location means that it has a wide footprint, but by the same token it has to be relatively large, weighing perhaps 30 000 kg. The main GEO operator, the International Maritime Satellite Organisation (Inmarsat), is currently capable of transmitting at 64 Kbps via a laptop-sized terminal.

Although deregulated in the UK in 1992, satellite communications were held back in Europe by a reluctance to deregulate the sector in the face of objections by incumbent fixed-wire carriers. The key to progress was the creation of non-geostationary satellite systems for voice and data services that would function with mobile handsets or laptops no matter where in the world such devices were carried.

In order to span the entire world, a network of satellites was needed at what would need to be an enormous cost, with the number of satellites depending on their height above the earth – the higher, the fewer. The consortia which set out to build new networks adopted either low earth orbit (LEO) or medium earth orbit (MEO) networks. These LEOs were divided between 'little' LEOs, designed for positioning and mobile messaging and weighing between 50 and 100 kg, and 'big' LEOs designed for fixed and mobile telephony and data and weighing between 450 and 700 kg. The LEOs, which were to operate in the 1.5–2.5 GHz L-band, would provide better transmission quality, but wear out more quickly, compared to 'big' MEOs which typically weighed between 2500 and 3000 kg – their lifetime was between 5 and 7 years compared to 12 years for MEOs – and cost more to operate.

By November 1995, there were seven companies planning to provide services via LEOs, including newcomers Constellation and Ellipso.[19] ICO was the only non-American-dominated consortium. All planned to sell to 'wholesalers' which would distribute their capacity throughout the world. These would mostly be carriers such as Vodafone which would distribute for GlobalStar wherever it operated. Where appropriate, calls would be routed via terrestrial GSM networks, which would provide a major incentive to use GSM as a common global standard. The problem was that terrestrial GSM services became so cheap, and their range so wide as a consequence of roaming agreements, that satellite-based systems appeared to have lost much of their key sales advantage, namely their 'use anywhere' characteristic.[20] This badly affected the four networks, discussed below, that initially got off the ground.

Compared to terrestrial links, these satellite systems have certain defects. First, it is difficult to maintain the same quality for voice telephony, especially in poor signal conditions. Second, the speed of data transmission is even slower than GSM, and hence incapable of transferring, say, video images. The capability of the uplink, which is narrowband because it operates on low power, is even more limited than the downlink. These difficulties are related to the spectrum in use and the distance of satellites from the earth.

Handsets for these satellite systems are generally more bulky because they have to be capable of sending signals via both satellites and earth stations. Furthermore, for now, the antenna must at least reach the top of the user's head to avoid blocking the signal, and has to be outdoors to provide a clear line of sight to a reasonable proportion of the sky.[21] While handsets could be made more powerful, this would increase battery consumption and present additional health risks. Improved antenna efficiency awaits technological developments.

Despite the high overall cost of a LEO or MEO network compared to a GEO network, the former do have clear advantages including much shorter signal delays, the ability to save costs per launch by recourse to launches of multiple, smaller satellites, and the relative ease of upgrading network capacity.

It must also be borne in mind, however, that the drawbacks of any satellite system need to be considered in relation to the objectives of the network. In countries where terrestrial links of any kind are limited, even a modest satellite service largely confined to voice telephony may contribute enormously to overall communication needs. In contrast, a businessman is likely to be able to achieve superior connectivity via other means in the vast majority of places visited.

Limited bandwidth networks

Although there were others – such as little LEO operator Orbcomm which eventually filed for Chapter 11 bankruptcy in September 2001[22] – the networks discussed below have tended to dominate discussion when assessing prospects for this market.

Iridium

Iridium was originally proposed as a network requiring 66 small satellites in low earth orbit and 15 'gateway' earth stations at a cost of $5 billion. It was to use the GSM standard. The main partner was Motorola. A number of European partners such as Telecom Italia and Veba participated by way of partnership agreements and/or investments in the US-dominated consortium.[23] It received FCC approval in January 1995. Iridium was notified to the European Commission in August 1995, and was duly approved.[24] The first satellite was put in orbit in January 1997, and the official launch date was set as September 1998. However, service was not commenced until November shortly after the technology was approved by international regulators, and Iridium suffered a further setback in December when one of its two handset suppliers, Kyocera of Japan, was unable to deliver for technical reasons, causing its UK agent, Orange, to hold off selling the service until February 1999.

Services provided were voice, paging and data. Its strategy was not to compete with terrestrial mobile operators but rather to use them as its agents, and over 200 operators in 80 countries signed agreements. The handsets were larger and heavier than the then current mobiles and sold initially for roughly $3000. In addition, call charges were high. Motorola, with its 20 per cent stake, was determined that the company should not collapse under the strain of failing to meet banking covenants in February 1999, at which point Iridium's shares were worth $27, having fallen from a peak of $71 in May 1998, but even by the end of March subscribers numbered only 10 000 compared to projections of 30 000. By May, the share price had fallen to $7, and the covenants were subsequently waived until August, whereupon Iridium defaulted on $800 million of debt. With $1 billion still needed to complete the network, and Motorola refusing to provide further finance unless the other partners chipped in, the share price fell to just over $3 prior to its suspension as rumours circulated of a Chapter 11 bankruptcy filing in the USA that duly came about in September.[25]

The then DDI of Japan, one of Iridium's largest investors with an exposure equivalent to 10 per cent of the outstanding equity, offered to inject further funds if the business plan was altered. It noted in particular that GSM had become so widespread that targeting international businessmen was no longer realistic. Despite this, in February 2000, DDI announced its intention to liquidate Nippon Iridium, although prospects improved when Craig McCaw invested $75 million in Iridium to provide sufficient funds to keep it going until mid-2000. He also invested in Intermediate Circular Orbit (ICO) with a view to integrating both companies with the Teledesic project (see section on ICO Global Communications). However, he subsequently decided to concentrate on ICO and only the privately-owned US company Crescent Communications expressed any interest in taking over Iridium's assets. Although merchant bank Castle Harlan and Venture Partners also looked over Iridium, the absence of a specific proposal meant that the network of satellites was expected to be progressively shut down over a six-month period and allowed to burn up in a low orbit.

In November 2000, a new company trading as Iridium Satellite LLC (ISLLC), backed by airline veteran Dan Colussy and with Boeing replacing Motorola, was set up to buy Iridium's assets out of bankruptcy for

$25 million, of which only $6.5 million was in cash, with a view to providing services to companies needing satellite communications. Other potential purchasers, at least five in number, promptly claimed that Colussy had conspired with Motorola to buy Iridium for far less than they were prepared to offer. In December, the US Defense Department agreed to purchase three years of telephony services from ISLLC, but the estimated 70 satellites operational at this time were due to re-enter the earth's atmosphere, commencing in 2005, raising question marks over Iridium's longer-term future.

In March 2001, ISLLC announced that it would launch commercial voice services in April and that it had signed up 13 non-exclusive service providers to offer worldwide distribution capabilities for industry and governments. The launch of data services eventually transpired in June providing Internet access at a maximum – and for most potential users inadequate – 10 Kbps. By August, the number of service providers had grown to 16 and included Global Plus with distribution in 35 African countries.

GlobalStar

GlobalStar was originally proposed as a network requiring 52 LEO satellites (48 active, 4 spares) of which the first was launched in February 1998 and the last early in 2000. It was developed by Alcatel of France, Loral (the US defence group with a 45 per cent stake) and Qualcomm. Other partners included Aerospatiale, Alenia and Deutsche Aerospace.[26] All service partners were obliged to take an equity stake in the project and hence it was fully funded.

In September 1998, a rocket carrying twelve of its satellites crashed shortly after launch, and although this did not overly delay the project, it had to commence services without its full complement of satellites. Handsets initially weighed 350 g and were priced between $880 and $1400, and call rates were set at a maximum of $1.50 per minute for national traffic and $2.99 a minute worldwide. GlobalStar claimed that it would not face the same financial problems as Iridium since its investors were all substantial telecommunications companies and its business model was to offer services in regions where the alternative was no service at all. For example, it supplied a service via Vodafone covering rural Australia and via Iusacell covering rural Mexico.

GlobalStar used the CDMA standard and limited services commenced in September 1999, although there was to be a gradual regional roll out and some areas with minimal population would not be covered. Subscribers were initially scarce on the ground, but it was hoped that numbers would pick up once data could be sent at 9.6 Kbps. In July 2000, it cut the prices of its Qualcomm handsets and reduced the minimum per minute charge to less than $1, but its share price nevertheless fell to $11 compared to $47 at the beginning of 2000 – not surprisingly given a mere 13 000 customers. Further equity funding was subsequently provided by its partners and the likes of Bear Stearns and ChinaSat, but within the space of a month, in October 2000, Globalstar was forced to admit that its revenues were negligible and that the business model requiring telco agents to sell its services in their domestic markets was a failure. Its share price fell to $2.

In January 2001, faced with the need, but not the ability, to service debts of $1.5 billion, Globalstar suspended interest payments. Its customer base at the time stood at 31 200. In April, it recorded a net loss of $3.8 billion for the full year, defaulted on its debt, took three satellites out of service, admitted that it was being sued for violating agreements and misleading investors – and vowed to restructure and soldier on. Its share price fell to $0.43. The inevitable Chapter 11 bankruptcy could only be postponed until February 2002.

ICO Global Communications

ICO Global Communications, was affiliated to Inmarsat which already provided voice, data and tracking services via satellites in medium earth orbit (MEOs) to maritime and aeronautical markets at a relatively high price. It required 12 satellites (ten active and two spares) in the highest orbit at a forecast cost of $4.7 billion. Unlike Iridium, which relied almost entirely upon inter-satellite signal transmission, and Globalstar, which constructed more than 100 expensive gateways, ICO set out to achieve a better compromise with 12 gateways overall. The members of this consortium included Telefónica, Sonera, Swisscom and T-Mobil. In October 1997, BT became its fiftieth shareholder, and its numbers subsequently exceeded ninety. It used the GSM standard. It was approved by the European Commission at the end of 1996.[27] The proposed start-up date was 2000. Handsets were to be priced at roughly $1000, but it did at least own its own dedicated fibre-optic network, and hence its call charges were expected to range between $0.5 and $3 a minute.

ICO made a public issue of shares in 1998 priced at $12, which raised $120 million rather than the expected $500 million. However, a rights issue in April 1999 had to be launched at only $5 because of the fallout from Iridium's financial disarray,[28] and it was struggling to obtain further finance from its members. A $1.4 billion rescue package involving the issue of new 'B' shares, which reduced the voting rights of existing

'A' shareholders, ran into difficulties in August 1999 and the company was forced, like Iridium, to seek Chapter 11 protection. In October 1999, $225 million was raised from key shareholders and optimistic noises were made about the remaining $1 billion needed to emerge from Chapter 11. It was thought possible to salvage the company because the earth stations were already built even though the satellites had not been launched.

In November, Craig McCaw, one of the founders of Teledesic, a 288-satellite network set up in1994 to provide high-speed Internet access which also boasted Bill Gates, Prince Alwaleed Bin Talal and Boeing among its investors (see section on Teledesic), announced plans for a group of investors led by himself to raise the needed funds. Shortly after, Subhash Chandra, owner of India's largest media company, Zee Telefilms, offered to provide the money in return for a 74 per cent stake. However, it was McCaw who emerged with New ICO after paying out $1.2 billion in May 2000. He was particularly happy as he had paid $1.2 billion for a company in which $3 billion had been invested. The intention was to transfer McCaw's satellite assets into a holding company called ICO-Teledesic Global (ICO-TG) which was to have New ICO and Teledesic as its wholly-owned subsidiaries.

The first satellite, launched in March 2000, crashed into the ocean. As it was fully insured, this was thought to be a positive omen. New satellites were to be launched, but whereas services were expected to start in October 2002, they were set back to 2003 in order to improve signal quality and provide data rates of 144 Kbps. The target market is the corporate sector, with rural and remote areas a secondary consideration.

Odyssey

Odyssey was originally proposed as a network requiring 12 big MEOs. The main partners were TRW, the US aerospace manufacturer, and Teleglobe, a Canadian international carrier. The proposed start-up date was 1999. However, in December 1997, TRW announced that it would be taking a 7 per cent stake in ICO, thereby terminating the Odyssey system.

Reappraisal of the networks

Although narrowband services, including voice, data, fax and paging, are currently available, the weakness in the underlying business model has been demonstrated only too clearly by the demise of Odyssey and Iridium as well as ICO's and Globalstar's brush with Chapter 11 – all briefly noted above. Furthermore, prospects for the survivors among them will be affected by similar newcomers such as Ellipso, a 17-satellite network where the prime contractor and 10 per cent stakeholder in the relatively cheap $1.5 billion project is Boeing, the target launch date is 2002 and the target markets are densely populated areas. However, the prospects for narrowband services also depend upon broadband (up to 2 Mbps) projects aimed primarily at the laptop market such as those being pursued by Inmarsat, Teledesic and Skybridge.[29]

Inmarsat

Inmarsat, the privatised (in 1999) former intergovernmental organisation supplying services to ships, is a subsidiary of Inmarsat Ventures and has as its main shareholders Telenor Satellite Services with a 15 per cent stake and Lockheed Martin Global with 14 per cent. It had nine GEO satellites in orbit with five able to transmit at 3G speeds when, in June 2001, it placed an order for three further satellites worth $1.7 billion as part of its planned Inmarsat 4 network which was expected to start operations in 2004, transmitting to over 80 per cent of the earth's landmass at up to 432 Kbps. The projected customer base over an eight-year period is 600 000. In the meantime, its regionally based Broadband Global Area Network Services (B-GANS), operating for now as a form of mobile ISDN at 64 Kbps, with charges averaging $7.50 a minute and terminals costing $11 000, would provide services at 144 Kbps from the end of 2002.

Teledesic

Strictly a non-geostationary fixed satellite service (NGSO FSS) system, but widely known as the 'Internet-in-the-sky', Teledesic originally involved 288 LEOs transmitting mainly on the 1500 MHz band at a cost of $9 billion. It was thought unlikely to leave the ground until Boeing agreed to increase its commitment, but with the approval of the FCC it pencilled in a launch date of 2003 with satellite dishes initially set to cost $2000 (and hence intended primarily for the corporate market). It was proposed that two-way data would be sent at 2 Mbps.

Teledesic was initially obliged to share spectrum with Celestri, backed by Motorola. However, in May 1998, Motorola abandoned Celestri and took a 26 per cent stake in Teledesic in return for a combination of cash and design and development work on that project. The stakes held by Craig McCaw, Bill Gates and Prince Alwaleed Bin Tahal were adjusted as a consequence, and with Motorola at a more advanced state of satellite construction, the project received a considerable impetus.

In March 2001, plans to fold New ICO and Teledesic into ICO-TG were shelved; the first New ICO satellite launch was pencilled in for

June, to be followed by 11 more of which two would be spares; and a joint venture between ICO-TG and MEO operator Ellipso was announced possibly leading to the takeover of the latter. A second joint venture, between ICO-TG and Netherlands-based CCI International, whose plans had been put on hold in the aftermath of the Iridium bankruptcy, was announced at the same time.

Progress with Teledesic itself was slow. Only in February 2002 did it announce that it had contracted Alenia Spazio to build the first two of the proposed 30 satellites needed for its network which would be altered from a LEO to a MEO network. It anticipated that the first 12 would cost $1 billion to build. Service launch was indicated as sometime in 2005.

SkyBridge

Despite Celestri's demise, the SkyBridge project (otherwise known as CyberStar and led by Alcatel with the support of the likes of Loral, Qualcomm, Mitsubishi and Toshiba), which was initially set up to utilise 64 satellites and operate on the 1 GHz band, shared with existing satellites, also offered potentially tough competition. In June 1998, it was announced that the project would be upgraded to 80 satellites in order to increase its capacity by 50 per cent, with the total cost rising by $700 million to $4.2 billion. New investors in the form of carriers were expected to provide the additional funds.

Conclusions

Needless to say, there have been many hurdles to be overcome by all types of satellite network including, for example, the negotiation of earth station rights in less-developed countries, the allocation of frequencies[30] and the clarification of relationships between satellite and terrestrial services. At one stage, for example, the US authorities provisionally imposed a $1 billion charge on ICO to compensate US users of radio frequencies which needed to be re-allocated to ICO. There have also been ongoing concerns over the dominant role played by US companies in the consortia.

Expensive new satellite services were always perceived as somewhat risky,[31] but the reality is that their prospects have declined at an astonishing rate. It is less than a decade since such services were widely seen as the obvious answer to meeting the problem of 'go-anywhere' communications. However, in the space of a few years the reduced size, weight and cost of cellular handsets and the implementation of global standards for cellular networks, combined with a big reduction in wireless tariffs, have combined to destroy much of the rationale for satellite provision. This conclusion is very strongly held in relation to voice telephony whereas prospects for broadband networks, when they eventually come on stream, remain for now the subject of considerable disagreement. In part, this reflects the difficulties faced by terrestrial suppliers of broadband services via Asymmetric Digital Subscriber Line (ADSL) or cable modems since such services are currently uneconomic to supply outside major conurbations[32] and, in recent times, several DSL operators have gone into bankruptcy.[33]

Although it is possible to keep costs down by opting for geostationary networks, it has to be borne in mind that most of the prime locations – able to send a strong signal to high-income areas in the USA and Europe – are already occupied, and that any new slot must be separated by at least two degrees from any pre-existing slots.[34] Furthermore, satellites cannot be acquired off the shelf. Given the need for something close to technical perfection once in orbit, and in the launch itself, satellites need to be custom built[35] and exhaustively tested before being launched. This implies that existing geostationary satellite networks will remain profitable[36] because it will probably never be economic to roll out fibre-optic connections to large parts of the world's surface. In effect, the trade-off between satellite (limited bandwidth but limitless reach) and fibre-optics (unlimited bandwidth but limited reach) will never be fully resolved in favour of one alternative.

Hopefully, some lessons will have been learned from earlier failures. For example, there are the obvious (in retrospect at least) difficulties associated with signals that require line of sight and signal distortions resulting from excessive distance. In business terms, the most obvious lesson is not to incur excessive levels of debt, especially since that results in the temptation to try and recover costs as soon as possible via prices higher than the market will bear. In this respect, the bankrupt satellite operators appeared not to have learned much from the experience of their terrestrial cellular rivals in relation to the price elasticity of demand for mobile communication – that is, that as prices fall demand rises disproportionately yielding increasing revenues.

However, the waters are likely to be muddied somewhat by the advent of 3G since in some cases the licence costs have run into billions of dollars, and network roll-outs will possibly cost similar sums. Seen in relation to such numbers, the seemingly astronomical, and unrecoverable, costs of the failed LEOs and MEOs no longer seem quite so implausible. Hence, if, for example, there is a public outcry over, say, the need to build huge numbers of new towers for terrestrial networks, satellite provision may come to be viewed more favourably.[37]

Wireless application protocol (WAP)

WAP was originally developed by Openwave and its use is currently controlled by the WAP Forum, consisting of over 200 bodies involved in some way with upgrading the capabilities of hand-held wireless communications, whether handsets, pagers, PDAs or other types of terminal. The WAP Forum was set up in June 1997 by the likes of Ericsson, Motorola and Nokia.

WAP is based on a number of Internet standards such as the Internet Protocol (IP) – in other words it is not a new technology as such. WAP presents Internet content in a special text format for mobile handsets using the WML, a stripped-down version of HTML – often known as 'micro-browser' software. By implication, WAP-accessible websites must have their content generated both in HTML and WML. WAP 1.0 is designed to function best within the constraints of modest processing power, a small display area and an alphanumeric keypad – hence Web pages need to be stripped of animation and graphics and adjustments need to be made to allow for the low pixel count of the typical handset screen. It will work with any kind of air interface associated with 2G technology and can cope with any low-bandwidth application such as circuit-switched data and SMS.

WAP technology is relatively unsophisticated. Because the handset does not have a lot of in-built processing power, WAP requires that the network intelligence is placed not in the handset but at a gateway that links the wireless network to the fixed-wire Internet. Intelligence installed at the gateway provides protocol translation and optimises the transfer of data to and from the terminal.

The main complaint when WAP was launched was that, being a dial-up system, it operated so slowly that accessing anything other than the simplest form of textual information was an agonising experience, especially for anyone used to a PC. Further, the screen size was very off-putting given how little information could be viewed at any one time. Although this was not thought to be too much of a deterrent where accessing the likes of stock market data, timetables or banking was concerned – although the latter raised issues of security – the initial problems such as broken connections and the cost of time waiting to make connections to the Internet tended to discourage potential users. Also, there was little incentive for handset manufacturers to improve their display capability just for low-bandwidth applications. A final issue was that tapping the keys for even a limited search was a potentially lengthy business and it was easy to make mistakes – an issue partly addressed by the likes of Google Number Search which makes an intelligent guess at the intended word simply from a sequence of single key presses. Nevertheless, part of the problem was that potential users were given the mistaken impression that they could browse the Internet just as they could with their PC – a truly mobile Internet – and that was never the purpose of the exercise.

Given its problems, there has been a tendency in more recent times to argue that WAP will take off once it is able to support GPRS (and later UMTS) which will give it 'always-on' capability. The development of WAP 2.0 which is based on xHTML and capable of supporting EDGE, GPRS and UMTS while allowing for the use of a variety of additional input devices whether keyboards, touchscreens or stylii, is likely to improve WAP's acceptability.

General packet radio service (GPRS)

The general packet radio service (sometimes system) has been slow to establish itself, partly because of the somewhat adverse reaction to WAP. Also, it was always intended to be an intermediate technology, and it was initially unclear how long it would have the marketplace to itself before 3G took over. In practice, the reality has accordingly been that GPRS has come into its own as the prospects for 3G have receded somewhat – indeed, some network operators are no longer convinced that they wish to proceed beyond 2.5G – but the reluctance to invest heavily in an intermediate technology has meant that handset availability has been a problem in many countries. As is clear from Table 2.1 below, GPRS-enabled networks are now the norm in Western Europe, and are becoming commonplace elsewhere in the world where GSM is the 2G standard. Where CDMA is the 2G standard, the equivalent 2.5G progression is via cdma2000 1xRTT, the intermediate version of cdma2000 3xRTT.

The GPRS infrastructure and handsets support data transmission at up to 13.4 Kbps per channel, combining several downlink channels with a single uplink channel. In some respects the building out of GPRS networks is the relatively easy part. Once a GSM network is up and running it is necessary only to add two nodes in the core network (Serving GPRS Support Node and Gateway GPRS Support Node) and to add a circuit board to the radio network to allow packets to be transported.[38] The issues that then need to be resolved include maximising the stability of the new network, making arrangements for roaming and developing applications and services. It is also obviously necessary to have sufficient capacity to meet the demand for GPRS-enabled handsets.

Table 2.1 Progress with GPRS

Country	Operator	*Contracted date* Launch date[1]	Equipment supplier
Western Europe			
Austria	ONE	February 2001	Nokia
	Max.mobil	April 2001	Siemens
	Mobilkom	August 2000	Cisco/Motorola/ Nortel
	Tele.ring	January 2001	Alcatel
Belgium	KPN Orange	Early 2002	Ericsson
	Mobistar	March 2001	Nokia/Nortel
	Proximus	July 2001	Motorola/Nokia
Denmark	Orange (Mobilix)	January 2001	Nokia
	Sonofon	December 2000	Nokia
	TDC	January 2001	Ericsson/Nokia
	Telia	February 2002	Ericsson
Finland	Ålands Mobiltelefon	n/a	Ericsson
	Radiolinja	September 2001	Nokia
	Sonera	December 2000	Ericsson/Nokia
	Suomen 2G	n/a	Ericsson
	Telia Finland	October 2001	Nokia/Siemens
France	Bouygues Télécom	January 2002	Cisco/Ericsson/ Nortel
	Orange	February 2002	Alcatel/Ericsson/ Motorola/Nortel
	SFR	March 2002	Alcatel/Nokia/ Siemens
Germany	Vodafone D2	February 2001	Ericsson/Siemens
	E-Plus	March 2001	Nokia
	T-Mobile	June 2000	Alcatel/Cisco/ Ericsson/Motorola/ Siemens
	Viag Interkom	January 2001	Nokia/Nortel
Greece	CosmOTE	January 2001	Nokia
	Vodafone	March 2001	Ericsson/Nokia
	Tele Stet	July 2001	Ericsson
Iceland	Islandssimi	February 2001	Ericsson
	TAL	February 2001	Ericsson/Nortel
Ireland	Digifone	*June 2000*	Nortel
	Eircell	January 2002	Ericsson/Nortel
Italy	Blu	December 2000	Nokia/Nortel
	Omnitel	November 2000	Nokia
	TIM	October 2000	Ericsson/Siemens
	Wind	November 2000	Alcatel/Ericsson/ Siemens

Table 2.1 Continued

Country	**Operator**	***Contracted date*** **Launch date**[1]	**Equipment supplier**
Liechtenstein	Viag Europlattform	n/a	Nokia
Luxembourg	LuxGSM	February 2001	Siemens
	Tango	January 2001	Ericsson
Malta	Go Mobile	August 2001	Nortel
	Vodafone	August 2001	Siemens
Monaco	Monaco Telecom	n/a	Nortel
Netherlands	Ben	December 2001	Nokia
	Dutchtone	Q2 2002	Nokia/Nortel
	KPN Mobile	December 2000	Nokia
	Libertel	April 2001	Cisco/Ericsson
	Telfort	March 2001	Ericsson
Norway	NetCom GSM	January 2001	Nokia/Siemens
	Telenor	February 2001	Ericsson/Nokia
Portugal	Optimus	July 2001	Ericsson/ Motorola/Nokia
	TMN	November 2000	Alcatel/Siemens
	Vodafone	April 2001	Ericsson
Spain	Amena	July 2001	Ericsson
	Telefónica	January 2001	Cisco/Ericsson/ Motorola/Nokia/ Nortel/Siemens
	Vodafone	November 2001	Ericsson/Nortel/ Siemens
Sweden	Tele2	October 2001	Cisco/Motorola/ Siemens
	Telia Mobitel	September 2001	Ericsson/Nokia
	Vodafone	December 2000	Nokia
Switzerland	Orange	September 2001	Nokia
	Swisscom	February 2002	Ericsson
	TDC Schweiz	September 2000	Nokia
UK	mmO_2	June 2000	Cisco/Motorola
	One-2-One	Q1 2002	Ericsson/Nortel
	Orange	February 2002	Ericsson
	Vodafone	April 2001	Ericsson
Rest of the World			
Argentina	Telecom Personal	*December 2000*	Ericsson
Australia	C&W Optus	*May 1999**	Nokia/Nortel
Australia	Telstra	March 2001	Ericsson/Nortel
Bahrain	Batelco	*November 2001*	Ericsson
Bolivia	Entel Móvil	*November 2000*	Ericsson
Bolivia	Nuevatel	*February 2001*	Nokia

Table 2.1 Continued

Country	Operator	*Contracted date* Launch date[1]	Equipment supplier
Brazil	Telemar	*June/July 2001*	Nokia/Alcatel
Brazil	Telesp	*July 2001*	Lucent/Motorola
Bulgaria	GloBul	*June 2001*	Motorola
Canada	Rogers Wireless	July 2001	Ericsson/Motorola
Canada	Microcell	June 2001	Ericsson
Chile	Entel	October 2001	Ericsson
China[2]	China Mobile	Various	Various
China	China Unicom	*November 1999**	Nokia
China	China Unicom	*September 2000**	Nortel
Croatia	CroNet	July 2001	Siemens
Croatia	Vip-Net	June 2001	Ericsson
Czech Republic	EuroTel	*October 2000**	Nokia
Egypt	MobiNil	*February 2001*	Alcatel/Nokia
Egypt	Orascom	*July 2001*	Alcatel
El Salvador	CTE Personal	*March 2001*	Alcatel
Estonia	EMT	July 2001	Ericsson
Estonia	Radiolinja Eesti	*May 2001*	Nokia
Georgia	Magticom	*February 2001*	Motorola
Hong Kong	New World Mobility	October 2001	Nokia
Hong Kong	PCCW	November 2000*	Nokia
Hong Kong	SmarTone	*July 2000**	Ericsson
Hong Kong	Sunday	*Feb/May 2001*	Nokia/Nortel
Hungary	Pannon	June 2001	Ericsson
Hungary	Vodafone	December 2001	Nokia
India	Bharti Enterprises	*October 2001*	Ericsson/ Motorola/Nokia
India	Bharti Enterprises	*December 2001*	Siemens
Indonesia	PT Satelindo	*November 2001*	Siemens
Indonesia	Telkomsel	*February 2002*	Motorola
Israel	Cellcom	*July 2001*	Nokia
Lebanon	France Télécom	April 2001	Ericsson
Lebanon	LibanCell	September 2001	Cisco/Motorola
Lithuania	UAB Omnitel	*October 2000**	Ericsson
Malaysia	DiGi Telecoms	*October 2000*	Ericsson/Motorola
Malaysia	TM Touch	*October 2001*	Alcatel/Motorola
Mexico	Telcel	*December 2000**	Ericsson
New Zealand	Vodafone	*August 2000**	Nokia
Philippines	Digitel	*September 2001*	Alcatel
Philippines	Globe Telecom	*November 1999**	Nokia
Philippines	SMART	*July 2000**	Nokia
Poland	Polkomtel	*July 1999**	Nokia
Poland	PTC	*August 2001*	Siemens
Poland	PTK Centertel	*October 2000**	Nokia
Qatar	Qatar Telecom	*June 2000**	Alcatel
Russia	VimpelCom	*January 2000**	Nokia

Table 2.1 Continued

Country	Operator	*Contracted date* Launch date[1]	Equipment supplier
Singapore	M1	October 2001	Nokia
Singapore	SingTel Mobile	*June 2000**	Ericsson
Slovakia	Globtel	*March 2001*	Nokia
Taiwan	Chunghwa Telecom	August 2001	Nokia/Nortel
Taiwan	Far EasTone	*December 1999**	Ericsson
Taiwan	KG Telecom	*September 2000**	Nokia
Thailand	AIS	September 2001	Nokia
Thailand	CP Orange	*May 2001*	Alcatel
Thailand	CP Orange	*September 2001*	Motorola
Thailand	TAC	November 2001	Nokia
Tunisia	Tunisie Telecom	*June 2000**	Ericsson/Alcatel
Turkey	Aycell	*January 2002*	Alcatel
Turkey	Turkcell	March 2001	Ericsson
Ukraine	Kyivstar	*January 2001*	Ericsson
Ukraine	UMC	*March 2001*	Siemens
UAE	Etisalat	January 2002	Alcatel
USA	AT&T Wireless[3]	July 2001	Nokia/Nortel
USA	Cingular Wireless[3]	August 2001	Nokia
USA	Omnipoint	*Feb/August 1999**	Ericsson/Nortel
USA	VoiceStream W'less	November 2001	Nortel

Notes: Information from separate sources is sometimes conflicting, and some countries have been excluded because of uncertainty about the entries. The list is as accurate as can be determined via cross-checking as of February 2002.

* now launched.

[1] It is difficult to be precise about launch dates because most operators begin with a limited business service and follow up with a mass-market service some months later. Also, launch dates are sometimes announced but nothing then happens until a new launch date is announced months later.

[2] There have been a large number of contracts. For example: (1) Fujian Mobile in one province. August 2000. Nokia (2) Guangdong Mobile. October 2000. Ericsson (3) Four provinces. *December 2000*. Nokia (4) Jiangsu Mobile. *March 2001*. Alcatel (5) Anhui and Shanghai regions. *March 2001*. Siemens (6) Nine provinces. *September 2001*. Ericsson (7) All 31 provinces. *October 2001*. Nokia (8) Seven municipalities/provinces/regions. *October 2001*. Motorola (9) Xinjiang province. *November 2001*. Siemens (10) Sichuan Mobile. *December 2001*. Ericsson.

[3] AT&T Wireless is expected to provide a full nationwide service in 2002 and Cingular Wireless in 2004. Cingular has a network sharing agreement with VoiceStream Wireless.

Sources: Web sites of equipment manufacturers; www.cellular-news.com; www.cit-online.com; www.totaltele.com; www.ft.com/fttelecoms.

According to Ghribi and Logrippo,[39] the main limitations of GPRS are first that transmission rates tend to be lower in practice than in theory since network operators are reluctant to allocate more than a few time slots to a single user, and, secondly, that there can be transit delays affecting quality of service.

Working out how to charge for packet-based services and to integrate customer relation management software with billing systems has been a major stumbling block. There is likely to be a fairly clear distinction emerging between the business market – primarily interested in checking e-mails and downloading data from corporate intranets – and the consumer market – concerned with surfing the Web and using WAP.

From a marketing perspective, the prevailing view is that users will not be overly concerned about the speed of data transmission as such. Rather, they will welcome the 'always-on' facility and the change to a packet-based charging system. However, there is as yet no clear pattern of prices, partly because operators are unsure how to compensate for the loss of voice channel capacity. There is little enthusiasm for flat-rate pricing since this would be expected to lead to excessive demands being placed upon available capacity. Given that the commonest pattern is, accordingly, to opt for a volume-based model with a certain amount of data bundled for free, there is a clear incentive to keep usage as close as possible to the amount of bundled data. In August 2001, huge variations were apparent between corporate prices charged within Western Europe.[40] TIM offered 60 MB (megabytes) for $32.24 per month plus $1.71 per additional 1 MB and BT offered 50 MB for $75, whereas Telefónica offered 100 MB for $333 and Max.mobil offered 100 MB for $359. Since consumers are not typically offered discounts, they pay even more. For example, the bundled price per megabyte (BPMB) ranged from $5 (Vodafone) to a remarkable $46 (Max.mobil). While mobility has value, it is hard to understand why data sent over a mobile line should cost up to 1000 per cent more than via a fixed-wire connection for the transport of 100 MB. According to the Yankee Group report, GPRS becomes uncompetitive once a limit of a meagre 500 kilobytes is reached, and also when 2 MB is reached if compared to an ADSL connection. For its part, the report concludes that a maximum charge of $2 per megabyte would be needed to build up GPRS into a mass-market commodity.

There are, however, some signs of emergent competition. For example, in Sweden, Telia announced in November 2001 that it would be offering all subscribers a free three-month trial of GPRS. Rival Tele2 promptly offered a free trial until May 2002. In Finland, Sonera announced in January 2002 that it would be scrapping its fixed monthly connection fee of $14.3 and would charge €0.01 (less than one cent) per kilobyte of data. The same method and roughly the same price was announced by Swisscom in February 2002.

A subsequent report by Analysys[41] predicted that there would be 5.9 million GPRS users in Western Europe by the end of 2002, rising to

110 million by the end of 2006, representing one-third of all cellular subscribers – roughly the proportion that other analysts reckon is needed for GPRS to take off. The authors argued that time-based charges would survive only as a short-term expedient, and that whereas volume-based pricing would be appropriate for basic network access, network operators would need to find a way to relate charges to the perceived value of the services provided. In practice, roughly 7.5 million GPRS-enabled handsets were sold in total during 2001, although this was widely regarded as a disappointing outcome and owning a GPRS-enabled handset does not necessarily mean that it is used for anything other than voice telephony. Whether the outlook will improve during 2002 is uncertain, but the main GSM handset manufacturers such as Nokia intend that virtually their entire range will be GPRS-enabled by the end of 2002.

In February 2002 at the 3GSM World Congress Samsung unveiled the first GPRS videophone handset capable of providing high-speed access to music videos, Internet broadcast, animation and other motion picture in colour. The viewing experience in terms of clarity and smoothness is determined by the bandwidth allocated to the transmission.

Cdma2000 1x

As previously noted, the variant of 2.5G used for CDMA networks is known as cdma2000 1xRTT which evolves into cdma2000 1xEV-DO and thence on to cdma2000 3xRTT. Because this progression is much less common that that via GPRS to W-CDMA, the individual country studies in Chapters 3 and 4 specify which operators have declared their intention to develop this route. At this juncture, therefore, it is appropriate merely to indicate broadly how this is developing.

By the end of 2001, almost 30 operators had indicated the likelihood that they would develop cdma2000 1xRTT, but relatively few had actually made real progress. In practice, there are three main locations for the technology, namely the USA, Asia and South America – it is irrelevant in Western Europe. In the former case, Verizon Wireless, Sprint PCS and the relatively small Leap Wireless and NTELOS either have introduced, or are already close to introducing, the technology, but even so it is fighting a fairly even battle with the adherents of W-CDMA. Meanwhile, in Canada, Bell Mobility launched a service in February 2002 with Telus hard on its heels. In Asia, the second-largest Japanese operator, KDDI, favours the technology and intends to introduce it this year, but is heavily outweighed by the W-CDMA adherents DoCoMo and J-Phone. Ironically – because the two initial 3G licences were for W-CDMA – South Korea is fast evolving into the home of cdma2000, driven by the need to have something reasonably fast available for the World Cup football tournament in May 2002. By the end of 2001, W-CDMA licensees SK Telecom and KTF were both well on the way to cdma2000 1xEV-DO, with cdma2000 licensee, the LG Telecom consortium, trailing behind. There is also interest in the likes of Thailand (CAT Wireless Multimedia) and Taiwan (Chunghwa Telecom). Meanwhile, in South America, the relative popularity of CDMA as 2G is developing into a reliance upon cdma2000 as in the cases of Brazil (Telesp Celular), Chile (Smartcom PCS) and Venezuela (Movilnet) all of which had at least commenced trials of cdma2000 1xRTT by the end of 2001.

I-Mode

Introduction

That i-mode, a play on the Japanese word for 'anywhere', is a success cannot be in doubt, as subscriber figures set out in Table 2.2 reveal. Commencing on 22 February 1999, it had acquired its 30 millionth customer at Xmas 2001 at which time the number of subscribers using the Java-enabled 503i handset, launched on 26 January 2001, reached 10 million. Although the quarterly growth rate has inevitably slowed, the absolute increase in the numbers signing up has stabilised over the past year at over five million per quarter. Figures on this scale indicate that i-mode has only ever been exceeded in terms of a consumer purchase by the Sony Walkman.

Table 2.2 I-mode subscribers[1] and growth rates by quarter: February 1999–June 2001

Date	Subscribers	Increase	Growth rate %
February 1999	0	—	—
June 1999	523 000	523 000	—
October 1999	2 235 000	1 712 000	327
February 2000	4 466 000	2 231 000	100
June 2000	8 290 000	3 824 000	86
October 2000	14 037 000	5 747 000	69
February 2001	19 777 000	5 740 000	41
June 2001	24 989 000	5 212 000	26

Note

1 08/08/1999 = 1 million; 15/03/2000 = 5 million; 06/08/2000 = 10 million; 22/11/2000 = 15 million; 04/03/2001 = 20 million; 01/07/2001 = 25 million; 24/12/2001 = 30 million.

Source: DoCoMo (adjusted).

According to DoCoMo in March 2002, the average user was spending roughly $61 a month, considerably more than the $50 for a voice only service and representing roughly an additional $4 billion of revenue, so it is hardly surprising that other mobile operators are wondering whether they should be copying DoCoMo, if only as a means to stem the decline in average revenue per user as 2G markets reach saturation. However, that is not necessarily as straightforward as it seems, and the reasons for this are discussed in the sections below.

Why i-mode has been a success

Predictably, there is no single reason for the rapid and enormous take-up rate of i-mode. According to DoCoMo itself, there are a number of crucial propositions, namely:

- I-mode is a packet-based rather than circuit-switched system, and hence provides always-on (rather than dial-up) connections merely by pressing the i-mode button on a conventional type of handset which is neither materially heavier – some weigh as little as 60 g – nor more difficult to use compared to voice-only handsets but which has a larger than normal LCD screen capable of showing six lines of 20 characters with a high level of resolution.
- Access to i-mode services can be done while on the move, and hence is ideally suited to those wishing to while away their journeys to work and so on.
- The use of compact HTML (cHTML) as the standard for writing software (rather than the WAP standard adopted in the EU) makes it easy for existing websites to reconfigure for presentation on i-mode handsets. It is also relatively easy to provide colour and animation.
- The basic subscription, at roughly $2.50 a month, is cheap, specimen material can be viewed for free and there are a huge range of premium services available for a further fee which can be added to the mobile phone bill.
- The range of content is not merely very extensive currently, but was already extensive at the time when i-mode was first launched.

To these can be added the following:

- Internet access via computers in the home is limited by western standards, partly because space in Japanese homes is at a premium. For certain groups, therefore, such as women and the young, who are also deprived of internet access at work, i-mode represents the most accessible means of internet connectivity.

- Japanese managers have often been unwilling to operate computers, regarding this as work fit only for secretaries.
- The Japanese are willing buyers of high-tech gadgets.
- i-mode handsets are regarded as good value. They are subsidised down to as little as $75.
- Modest subscription costs and charges based upon packets sent rather than time make i-mode services appear to be affordable.
- Many Japanese market segments, especially the youth market, are avid consumers of 'fun' services.
- The use of animation, colour and sound makes sites much more fun to use.
- Services can be 'personalised' with user-tailored menus.
- The ease of writing software in HTML, together with the ever-rising demand for services, provides the incentive to develop ever more sites – almost 50 000 so far – which in turn stimulates demand.
- Official partner sites have a guaranteed income source from shared subscription fees.

Initially, i-mode was aimed primarily at the consumer market, with a concentration upon entertainment and information and a lesser emphasis upon databases and on-line transactions. Roughly three-quarters of all subscribers use the e-mail facility, and other very popular services include the downloading of ring tones – facilitated by a built-in sound mixer – and of images such as photographs, transport timetables and mobile banking. Increasingly, the emphasis is being placed upon business transactions, particularly dealing in shares and airline bookings. The balance is now roughly even between ring-tone downloading (9 per cent of menu sites visited) plus games (11.5 per cent), entertainment (17 per cent) and information (13.5 per cent), and transactions (40.5 per cent) plus databases (8.5 per cent).

Sources of revenue

As indicated, DoCoMo earns a significant part of its revenues from the monthly subscription and charges for the transmission of data packets. The monthly fee of roughly $2.50 has to be added to a data charge of roughly $0.02 per kilobyte. These currently generate roughly $15 per subscriber per month. In addition, premium services yield a further $2 per month, and small sums are derived from service provider promotions that require interested parties to respond via their handsets.

Although there are almost 50 000 sites available, the 900 or so that are official are entitled to charge subscription fees ranging up to $2.50 per month. These fees are collected by DoCoMo which deducts a 9 per cent commission before passing the rest over to the originating site.[42] A surprisingly high proportion, around one-half, of all i-mode users subscribe to at least one such site. Significantly, i-mode subscribers also typically spend 15 per cent more on voice calls than other subscribers and, predictably, are much less subject to operator churn.

In the latter context it is notable that DoCoMo's handsets are operator-specific, with the downside that additional phones are needed when switching from one network to another. However, the offsetting benefit has been DoCoMo's ability to dictate the features that are thought to be desirable, such as colour screens. Because service providers know that these features will be available, they can safely set out to write software which relies heavily upon them to attract users' attention, and the availability of such features makes users willing in their turn to pay more than they would otherwise.

Will i-mode travel?[43]

On the face of it, the success of the i-mode model appears to be based upon the kind of sound business principles that ought to travel well. It is packet-based, service provider friendly, user friendly, reasonably cheap, offers a plethora of useful content and can be implemented without the need for expensive new billing systems and encryption coding. The technology works well without being intrusive, so the emphasis can be placed upon the needs of the customer. In effect, i-mode is a wireless Internet model – although this term is somewhat inappropriate given the clear distinction between i-mode and W-CDMA[44] – that actually operates profitably even though it has to use a fairly standard infrastructure with access at a fairly slow 9.6 Kbps. A further point is that widespread familiarity with i-mode should facilitate the transition to W-CDMA which is currently being launched on a modest scale.

Despite the above, European and US telcos[45] have not rushed out to introduce i-mode equivalents into their markets. The main reasons put forward are that

- Fixed-wire internet access is already fairly high and growing. Furthermore, this is true of domestic as well as business use, with all family members increasingly familiar with computers. Internet access via computer will almost certainly remain the preferred option other than when people are on the move.
- A lesser proportion of disposable income is devoted to entertainment.
- Less time is spent waiting around to get somewhere or for something to happen.

- There is less enthusiasm for gadgets.
- There is a greater preference for spoken communication.

For these reasons, it is unusual to find analysts who support the view that i-mode's prospects can ever be as good outside Japan as within it. On the other hand, some find the above points somewhat unconvincing, arguing that

- There are far more subscribers to mobile networks than to ISPs.
- Mobility is a feature of all Western societies.
- Younger mobile users are always keen to utilise new services.
- The growing use of e-mail and text messaging has accustomed people to accessing written communications via a screen.
- Everyone spends some time waiting around with time to kill.
- The substantial sums already expended on mobile phones suggests that people will pay for additional services provided they fulfil some kind of need.

Conclusions

At the end of the day, i-mode has little future outside Japan unless mobile network operators are willing to develop it as an alternative to WAP, and so far that means only KPN Mobile in which DoCoMo has taken a stake.[46] For most others, the issue is much more about the route to be taken in the evolution towards 3G, which generally involves GPRS. It is agreed that the crucial error was to market WAP as a substitute method of access to the Internet which it was incapable of delivering satisfactorily to users accustomed to the quality of access via their computers. DoCoMo preferred to emphasise the services to be provided rather than the technology, and hence was careful not to raise expectations unnecessarily. Hence, by and large, subscribers are happy with what is available. Since GPRS is also a packet-based system, it should overcome some of the obvious faults with WAP, and in any event it can be argued that as WAP develops it will provide the capability to 'push' data of local interest, such as weather conditions, to subscribers for whom such information has value. There is also a widespread desire for technology to evolve along a path determined by agreed standards, and while WAP is not yet established as *the* standard for mobile internet access outside Japan, it, or something similar, is expected to perform that role in the future. Hence, while there is some support for rolling up WAP and cHTML into an intermediate standard on the route to XML, this is far from achieving widespread acceptance.[47]

Smartphones and personal digital assistants (PDAs)

In an ideal world, a mobile handset would have the functionality of a PC. In fact, the functionality of a standard handset is truly extraordinary, but consumers have become increasingly blasé about technology and fail to appreciate that a modern handset does indeed have most of the functionality of a PC built not all that long ago – albeit much less than that of more recent models. Nevertheless, although a standard handset can be only so 'smart' given, inter alia, its physical limitations, there is no reason why a hand-held device cannot be a good deal smarter.

A so-called smartphone must be able to achieve a number of tasks similar to those achievable by a PC, including the conveyance of voice, Internet access and Web browsing, the sending of e-mails and a personal information manager (PIM) for taking notes plus an address book. The smartness of a smartphone is determined by its operating system (OS), processing power, data storage, software applications and interfaces with other devices. Because these make huge demands on power, some kind of power management tool is also fairly essential. Given the storage demands of a large address book plus e-mails, a capacity of several megabytes of memory is very important. A smartphone is in the above sense essentially a mobile handset with add-ons whereas a PDA is essentially a handheld computer for sending mobile data with the ability to double up as a handset for voice.

The sorts of issues raised above are unlikely to hold back development. Indeed, Toshiba, for example, announced that it would be able to provide up to 20 MB of memory on a 1.8-inch hard disk drive as early as March 2002. A more thorny issue is how to input data. As indicated above, the standard alphanumeric handset keyboard is far from ideal, even though many teenagers are surprisingly agile with it. Equally, a stylus-based system has its limitations. For optimum speed a 'qwerty' keyboard is the answer, which can either be integral to the device – inevitably involving greater size and weight although they can be made collapsible to avoid the former drawback – or attachable as needed, and, if separate, can be snapped on, connected by a plug-in wire or connected via a wireless interface such as Bluetooth.

For ease of browsing the Internet, there is no substitute for an 'always-on' connection. This can be provided by a variety of operating systems such as Palm OS, Symbian EPOC or Windows CE (WinCE). Symbian was spun out of Psion in 1998 – leaving it with a 28 per cent stake – to licence the EPOC operating system. It is supported by major terminal manufacturers such as Motorola, Sony, Ericsson and Nokia (with a 21 per cent stake) of whom the latter two already have Symbian-driven smartphones on the market. Microsoft's situation is interesting because, coming late onto the scene, it found almost all the prospective smartphone manufacturers licensed to Symbian, and only Samsung has adopted its Windows powered Smartphone 2002 software. The capabilities of Microsoft's PDAs are linked to the relatively sophisticated Windows operating system, and hence the WinCE version is likely to be favoured for models towards the top-end of the market. The latest version of WinCE, the PocketPC, is achieving inroads into the US market with the launch of devices such as the Trium Mondo, a phone-enabled hand-held device made by Mitsubishi and Microsoft that can do conference calls, call direct from the address book and send and receive faxes, and a more 'phone-centric' operating system, called Stinger, has also been reasonably successful.

An alternative PDA-type device that is rapidly catching on in the USA is the Blackberry, made by Research in Motion, which has an embedded wireless modem designed to facilitate wireless e-mails. The modem can be GPRS-enabled for use in Europe. The new Treo devices launched by Handspring in February 2002, utilising Palm OS, are also likely to prove popular with their combined functions of hand-held computer, PDA and GSM phone – viewed as a 'mission-critical' application by its makers in helping to create a hybrid device, sometimes referred to as a 'communicator', that goes beyond the concept of the PDA, provided, of course, that potential users are willing to hold hand-helds up to their ears. Again, it is worth noting that Treo devices support HTML, cHTML, WML and XML – a sign of future developments.

At the end of 2001, Nokia, with its new Communicator, overtook Palm (with its new i705) and Compaq (with its Accompli) to become the leading vendor of mobile hand-helds in Europe, thereby raising Symbian's share of the OS market to almost 35 per cent, with PalmOS trailing at 30 per cent and WinCE at 21 per cent. However, both PalmOS and PocketPC 2002 are expected to fight back during 2002. At the end of the day, quarterly sales of roughly half a million hand-held devices are dwarfed for now by sales of conventional mobile handsets, but the onset of GPRS, let alone 3G, is likely to move the balance in favour of PDAs and similar devices, although, as indicated by the previous discussion, precisely what variant of hand-held terminal will win the war for market share is difficult to predict. A gradual move towards hand-helds may be influenced by the widespread removal of handset subsidies since the true cost of standard handsets is much higher than most users have reason to believe, and paying the extra for hand-held functionality may seem attractive, especially in the business market.

3

UMTS Licensing in Western Europe

Table 3.1 summarises the results of the licensing process in Western Europe, with the subsequent text examining the process in each country in detail.

Table 3.1 Third-generation mobile telephony licences in Western Europe

Country[1]	No. of licences	Method	Month awarded	Winners[2] & Equipment[3]	Fee paid $ million[4]
Austria	6	Beauty contest + auction	Nov 2000	Hutchison 3G[a,d,e,f]	98+22
				Mannesmann 3G[b]	98
				Max.mobil[e,f,g,h]	103 + 44
				Mobilkom Austria[b,g]	104 + 44
				ONE[a,b,f]	104
				3G Mobile	101
Belgium	4	Auction	Mar 2001	KPN Mobile 3G[b,c]	150
				Mobistar[a,b,f]	150
				Proximus[f,h]	150
				Not awarded	
Denmark	4	Auction	Sep 2001	Hi3G Denmark[a]	118
				Orange	118
				TDC	118
				Telia[e,h]	118
Finland	4	Beauty contest + fee	Mar 1999	Radiolinja[e,f,h]	Nominal
				Sonera[b,f]	Nominal
				Suomen 3G[b,f]	Nominal
				Telia Finland[f,h]	Nominal
France	4	Beauty contest + fee	May 2001	France Télécom[a,b,f]	*4 520*[7]
				SFR[a,e,f,g,h]	*4 520*[7]
				Not awarded	
				Not awarded	

Table 3.1 Continued

Country[1]	No. of licences	Method	Month awarded	Winners[2] & Equipment[3]	Fee paid $ million[4]
France	2	Allocated	May 2002	Bouygues *Not awarded*	550 + 1% revenue p.a.
Germany	6	Auction	Aug 2000	E-Plus Hutchison[a,b,c] Group 3G[c] Mannesmann[b,h] MobilCom M'media[b,g] T-Mobile[d,f,g,h] Viag Interkom[f]	7 620 + 50 7 630 + 50 7 650 + 50 7 600 + 50 7 690 + 50 7 670
Greece	4	Auction	Jul 2001	CosmOTE Panafon STET Hellas *Not awarded*	157 139 126
Ireland	4	Beauty contest + fee	Jun 2002	* * * *	
Isle of Man	1	Allocated	May 2001	Manx Telecom[e,h]	Nil
Italy	5	Beauty contest + auction	Oct 2000	H3G (Andala)[a,b,d,e,h] IPSE 2000 Omnitel[f,g] Telecom Italia[b,e,h] Wind[a,b,f,h]	2015 + 683 2030 + 683 2020 2010 2015
Liechtenstein	4	Allocated	Various	Tele2/Tango Viag Europlattform[f] *Considering*[5] *Rejected*[5]	Nil Nil
Luxembourg	4	Beauty contest + annual fee	May 2002	EPT Orange Tele2 Not awarded	0.2 % revenue p.a. (min. 189,000)
Malta	?	?	?	*	
Monaco	1	Allocated	Jun 2000	Monaco Telecom[e,h]	Nil
Netherlands	5	Auction	Jul 2000	Dutchtone[a,b,f] KPN Mobile[b,c] Libertel-Vodafone[b] Telfort[b] 3G-Blue	407 664 667 401 369
Norway	4	Beauty contest + fee + annual fee	Nov 2000	*Broadband Mobile*[6] NetCom GSM[e,f,h] Telenor[b,f] Tele2	22 22 22 22

Table 3.1 Continued

Country[1]	No. of licences	Method	Month awarded	Winners[2] & Equipment[3]	Fee paid $ million[4]
Portugal	4	Beauty contest + fee + annual fee	Dec 2000	ONI-Way[e,g,h]	86
				Optimus[a,b,f,g]	86
				Telecel[b,g]	86
				TMN[a,b,e,h]	86
Spain	4	Beauty contest + fee + annual fee	Mar 2000	Airtel[g]	112
				Amena[b,h]	112
				Telefónica[b,g]	112
				Xfera[b,g]	112
Sweden	4	Beauty contest + annual fee	Dec 2000	Europolitan[b,f]	Nominal
				Hi3G Access[a,b,d,f]	Nominal
				Orange Sverige[a,b,f]	Nominal
				Tele2[b]	Nominal
Switzerland	4	Auction	Dec 2000	Dspeed[b]	29
				Orange[f]	32
				Swisscom[b]	29
				Team 3G	29
UK	5	Auction	Apr 2000	BT3G[b,g,i]	6 100
				Hutchison 3G[a,d,e,f,g,h]	6 700
				One-2-One[e,f,g,h]	6 100
				Orange[f]	6 200
				Vodafone[d,f,j]	9 100

Notes

* not yet allocated

[1] Countries underlined are not members of the European Union.

[2] Some of the licensees changed ownership structure and/or name after receiving a licence – see text below.

[3] The equipment suppliers are as follows. It should be noted that some contracts are provisional and the value of the work awarded is not normally spread evenly among the contractors.

[a] Alcatel

[b] Ericsson

[c] Lucent

[d] Motorola

[e] NEC

[f] Nokia

[g] Nortel

[h] Siemens

[i] Cisco

[j] Panasonic (Matsushita)

The only companies to have declared which software platform will be used specifically in their handsets are Fujitsu and Siemens, which have opted for Symbian, a platform also supported by Ericsson, Matsushita, Motorola and Nokia which are co-owners together with Palm. However, Microsoft has announced that its Stinger platform were available from the end of 2001 and under trial with Deutsche Telekom, Telefónica and Vodafone.

[4] The second figure represents the cost of acquiring additional spectrum made available at the end of the first stage of the auction.
[5] See text.
[6] Broadband Mobile returned its licence in August 2001. It will be reissued at some unspecified time.
[7] Subsequently adjusted – see text.

Source: Compiled by author from reports in media.

Individual countries

Austria

Final decisions on procedures for the award of 20-year UMTS licences were announced in July 2000. Twelve frequency packages of 5 MHz paired (upstream/downstream) – a total of 120 MHz – were put on offer with bidders obliged to offer to buy either two or three packages. Thus, the intention was that either four or six licences would be auctioned. The auction, in November, which was based upon the rule book established for the German auction, was preceded by a 'beauty contest' based upon applicants' technical and economic ability to deliver the service, including their ability to cover at least 25 per cent of the population by the end of 2003 and 50 per cent by the end of 2005. The auction rules required incremental bids above the highest in the previous round to be initially at least 10 per cent higher but falling gradually to 2 per cent higher. The auction would terminate when no valid bids were made for any of the packages in the course of a single round.

The six successful qualifiers for licences comprised 2G incumbents ONE (owned by E.ON 50.1 per cent, Telenor 17.45 per cent, Orange 17.45 per cent and TDC 15.0 per cent), Max.mobil (owned by Deutsche Telekom), Mannesmann 3G Mobilfunk (representing 2G licensee Tele.ring, majority owned by Vodafone) and Mobilkom Austria (Telecom Italia 25 per cent + 1 share), plus new entrants Hutchison 3G Austria and 3G Mobile Telecommunications (now wholly-owned by Telefónica). The incumbents were expected to win licences, which meant that new entrants could have been shut out, although the fact that there was the same number of bidders as the maximum number of licences indicated that bidders could not be expected to compete too hard given the modest size of the potential market. If successful, new entrants would be authorised to interconnect with existing 2G networks for no more than four years once their own network reached 20 per cent of the population.

At the end of the first day and nine rounds, each of the six bidders was in control of two packages with the total bid amounting to $575 million.[48] Although occasional interest was shown in bidding for three packages, the auction ended abruptly after only 14 rounds with all six bidders acquiring two packages at prices varying between $98 million and $104 million. The total raised was a meagre $609 million (€705 million) – little more than the minimum reserve price.

After the licences were awarded, a secondary auction took place for five 5 MHz lots, each offered at a reserve price of 350 million schillings ($21.9 million; €25.4 million), with a limit of two for any qualifying bidder that had been successful in the first auction. Initially, only Max.mobil and Mobilkom entered bids, and it took only two rounds to complete the bidding with these two each acquiring two packages and Hutchison 3G the fifth. All packages were sold at the reserve price, so the total raised by the entire auction only just exceeded the minimum set by the government.

Max.mobil later announced that it would be building an additional 1000 base stations with a view to achieving coverage of 50 per cent of the population by 2003. In mid-December, Vodafone acquired a further 25.2 per cent stake in Tele.ring – a fixed-wire and mobile operator – from the Osterreichische Bundesbahnen and the Verbund, and subsequently acquired the remaining shares from Citykom Austria. However, in May 2001 it sold its entire stake to Western Wireless International of the USA.

In October, ONE announced that it had taken a majority stake in broadband wireless provider eWave with the aim of adding W-LAN services to its UMTS provision, especially in hotels and airports. eWave claimed that it was capable of providing services at 12 Mbps without running into problems of interference. ONE also announced that it would launch its network during the second half of 2002 provided handsets were available.

Belgium

The government intended to sell four 20-year UMTS licences, each with a reserve price of €150 ($140) million, via a multi-round auction commencing in March 2001. A bandwidth of 155 MHz was on offer, mainly divided into four lots of 15 MHz paired plus 5 MHz unpaired. Incumbents Proximus (Belgacom 75 per cent, Vodafone 25 per cent) and Mobistar (Orange 50.7 per cent) which between them controlled virtually the entire 2G market seemed bound to win licences. Licensees were obliged to meet a series of progressive annual population coverage targets ranging from 30 to 85 per cent during years three to six of the licence. KPN Orange (now wholly-owned by KPN Mobile (KPN Telecom 85 per cent, DoCoMo 15 per cent) and bidding as KPN Mobile 3G Belgium), the third incumbent, claimed that it should be given a rebate on the licence cost because its relatively new PCN network had been costly to licence and roll out and still had not shown a profit, yet it would have to be made available for roaming to new UMTS licensees. This claim was rejected. Subsequently, Suez Lyonnaise des Eaux (SLE – now Suez) and Telefónica arranged to bid via a 60/40 joint venture, ST3G. However, cable operator Telenet refused to participate on the grounds that the market would not sustain four companies. In February, ST3G withdrew, leaving only the three incumbent bidders, and these duly won their licences at the minimum price, or just above, in an auction lasting one hour.

In November 2001, the licensees threatened to sue both federal and regional governments unless they provided guidelines for the roll-out of their infrastructure. Because of fears about health, several regional governments were intent on blocking the construction of new towers. In February 2002, Proximus, Mobistar and KPN filed a joint request with the government to have the official launch date set back one year to September 2002. The regulator initially recommended a six-month delay, but the government subsequently agreed to set the launch date back to September 2002.

Denmark

Initially there was no clearly established strategy for the introduction of UMTS other than that there would be a 'beauty contest' for four 15-year licences which could be expected to go to the 2G incumbents – TDC (formerly Tele Danmark), Sonofon (Telenor 53.5 per cent, BellSouth 46.5 per cent), Orange (formerly Mobilix – Orange 53.6 per cent, national railways 14 per cent) and Telia – unless the government was keen to prevent this and major international telcos joined consortia to make bids. However, in April 2000, the government decided to hold an auction on 5 September which was subsequently specified as 'single-round sealed-bid' – the first time for such a method in Europe. The objective was to provide opportunities for aggressive new entrants and to prevent collusion. All licensees would pay the lowest price submitted by a successful bidder, with the minimum set at $58 million. Each licence would comprise 15 MHz paired plus 5 MHz unpaired on a national basis, with the possibility of a further 5 MHz for the highest bidder. 2G incumbents would not be obliged to provide a roaming facility to new entrants. After objections were raised by prospective bidders, the coverage requirement was reined back to 30 per cent by the end of 2004 plus 80 per cent (down from 90 per cent) by the end of 2008, and the up-front payment was reduced to 25 per cent with the rest payable over a ten-year period. Network sharing would be permitted, but only when 80 per cent coverage had been achieved.

In the event, Tele2 chose not to bid, preferring to act as a MVNO utilising Telia's network – the reverse of the arrangement in Sweden. In total, five bids were made, with Telenor bidding alone since its partner in Sonofon, BellSouth, did not want to become involved, probably because it intended to dispose of its stake in Sonofon. However, Telenor lost out to Hi3G Denmark (Hutchison Whampoa 60 per cent, Investor AB 40 per cent) and all successful licensees ended up paying $118 million, almost exactly twice the minimum price and rather more than had generally been predicted.

Finland

Applications for four UMTS licences were invited in January 1999 – the first EU country so to do. A 'beauty contest' combined with a nominal fee was used, rather than an auction, to select the winners. The available spectrum consisted of four lots of 15 MHz paired plus 5 MHz unpaired. Each network licence ran for 20 years and each renewable frequency licence for 10 years. New entrants were permitted to roam onto existing 2G networks. No specific requirements were laid down for population coverage. In March, licences were awarded to Sonera, Radiolinja (a subsidiary of fixed-wire Elisa Communications, itself a subsidiary of Helsingin Puhelin), Telia Finland (building the Pan-Nordic 3G network) and Suomen 3G (majority owned by 42 local operators in a consortium with Tele2 (formerly NetCom AB) and trading as DNA Finland) with a stipulation that base stations should be built by the beginning of 2002. Also in March, Suomen 3G – which shared 36 common members with Suomen 2G – awarded Ericsson the first contract for a turn-key system capable of delivering GSM and UMTS.

In November 2001, the first-ever 3G call was allegedly made between Sonera's network and that of DoCoMo in Japan.

France

Despite the huge sums raised from the auction of UMTS licences in the UK, the French government was determined to conduct a 'beauty contest' for the four licences on offer because the three incumbents, the restructured Orange (France Télécom), Cégétel subsidiary SFR (Vivendi Universal/Vodafone/BT) and Bouygues Télécom (Bouygues/Telecom

Italia et al), faced severe difficulties in raising the kind of sums offered in the UK and hence were vociferous in their condemnation of a UK-style auction. Furthermore, it had to be remembered that France Télécom could not easily issue new shares to finance an expensive licence because the state's shareholding could not be reduced below 50 per cent. Bouygues was especially antagonistic towards an auction as shareholders Veba, BNP Paribas and Telecom Italia were all expected to bail out if they were required to finance the licence.

Nevertheless, the government could not afford to ignore the potential income flow altogether, and it accordingly hatched a compromise whereby the chosen four would pay Ffr32.5 billion (roughly €5 billion/ $4.5 billion) apiece, half during the first two years of their 15-year licences and the rest over the residual 13 years. This was equivalent to a total net present value of €15 billion, considerably less than in the UK, but at the same time the licences ran for five fewer years so they were not exactly cheap either.

The 'beauty contest' would be based upon 14 criteria carrying a total of 500 points. Of these, the most important were the scope and speed of network deployment (100 points), the coherence and credibility of the project (100 points), the coherence and credibility of the business plan (75 points) and services on offer (50 points). Licence winners would have to cover 25 per cent of the population after two years in respect of voice services and 20 per cent in respect of data services, rising to 80 and 60 per cent respectively after eight years. New entrants would be permitted to roam over existing 2G networks.

Hutchison, KPN Mobile and DoCoMo joined together in order to make a single bid, but internal wrangling suggested it might not end up being tabled. Meanwhile, Suez Lyonnaise des Eaux (SLE) joined with Telefónica in a 60/40 holding company, ST3G, as in Belgium, and later had discussions to bring Rallye-Casino, Groupe Arnault and TIW into the consortium. Rallye-Casino duly took a 10 per cent stake, and Groupe Arnault a 6 per cent stake, in the SLE part of the consortium.

It was decided that the qualification process would be set back to allow the dust to settle from the auction in Germany, with applications to be submitted before the end of January 2001. However, in November 2000, Deutsche Telekom announced that it would not be participating because it had no existing 2G network, and KPN made it clear that its priority was to win a licence in Belgium, so the possibility of only four bidders began to loom large. Analysts made the point that if Deutsche Telekom did not expect to make a profit from a licence, SLE/Telefónica could hardly expect to do so in the absence of an existing 2G network, so it came as little surprise when ST3G withdrew in January 2001. The government's response was to state that it would carry on regardless.

Bouygues Télécom[a] had hoped to bid in conjunction with part-owner Telecom Italia, but this plan was forestalled when the latter declined to raise its stake in the operator – it subsequently sold its holding to Bouygues in February 2002 leaving Bouygues with a 64.5 per cent stake. The ability of Bouygues Télécom to launch a solo bid was brought into question by the adverse market reaction to growing indebtedness in the telecoms sector at the end of 2000, but it was expected that if the loan was taken on by its main shareholder, Bouygues, it would be able to raise the €8 billion it needed before the application deadline at the end of January 2001.

In retrospect, the decision to go for a fairly high fixed fee rather than an auction initially looked to be astute given the insignificant sums raised in Italy and Austria. On the other hand, the fee was viewed as excessive by potential bidders for the same reason and, in essence, incompatible with claims that a 'beauty contest' was the method for allocating licences. As a partial response, the government announced early in January 2001 that although it was unwilling to reduce the licence fee, it would make the tax position for licence winners less onerous. However, this was insufficient to forestall the last-minute withdrawal of Bouygues Télécom, with Bouygues arguing that it was more sensible to restrict its subsidiary to GPRS which, it claimed, would be able to support 90 per cent of multimedia services. Both Orange and SFR responded to the possibility of some kind of reduced rate for the other two licences at a later date – an incentive reflecting the fact that the European Commission required the French government to issue four licences – with a demand for a guarantee that any subsequent issue of licences would be on the same terms as the initial allocation. The government announced in April 2001 that it would wait to see how services developed around Europe and did not intend to re-offer the unallocated licences until at least May 2002, subsequent upon the Spring elections.

The allocation of licences to Orange (which scored 379 points) and SFR (410 points) was confirmed in late May. Orange promised 58 per cent coverage by May 2003 and 93 per cent by May 2006, while SFR promised 70 and 97 per cent respectively. Orange planned to launch in June 2002, and SFR three months earlier. At the time, the government stated that if it issued additional licences at a price below that paid by Orange and SFR, their own fees would be retrospectively lowered to the same amount. At the end of September, SFR announced that it had placed the $563 million initial payment of its licence fee into a blocked bank account to indicate its desire for 'constructive dialogue' about the overall cost of the fee. SFR argued that delays by equipment manufacturers, refusal of sites by local authorities and the problems of the financial markets needed to be taken into account, and that the value of the licence had fallen to only $1.1 billion. However, when the government threatened legal measures to enforce compliance, SFR backed down.

In mid-October, the government announced that it would be extending the licence period to 20 years and would be re-offering the two unallocated licences in December. Instead of the existing high one-off licence fee, all licensees would pay roughly $550 million in 2002 plus 1 per cent of 3G revenues (excluding handset sales) thereafter, with the total payable eventually approximating the original one-off fee. Tele2 of Sweden was the first company to express a desire to participate in the second 'beauty contest', with Bouygues following suit, but Suez refused to bid. The submission of bids was later put back to May 2002 with a decision promised before the end of September. DoCoMo subsequently offered to take a small stake in a successful bidder – it was thought to favour Bouygues with which a technical partnership had been created to examine the potential for i-mode in France.

Germany

The government set out to auction between four and six 20-year UMTS licences commencing in July 2000. A spectrum of 120 MHz was divided into 12 blocks of 5 MHz paired, and each licence could contain either two or three blocks. Once a bidder had issued a bid for only two blocks, that bidder would not be permitted to issue a subsequent offer for three blocks. Bidding was to be started at roughly $90 million (€100 million) per licence, but licences were expected to cost many times that figure with a total outlay based on the UK experience expected to be in the region of $45 billion.[49] The successful bidders in the first stage would be permitted to bid in a secondary auction for five additional 1 MHz unpaired blocks plus any paired spectrum not sold during the initial auction (technically making it possible for a bidder to end up with four blocks). Unlike in the UK, the licences did not contain specified spectrum, with the regulator reserving the right to assign the spectrum at the end of the auction. Licensees were required to cover at least 25 per cent of the population by December 2003 and 50 per cent within three to five years from launch. Unusually, the government did not announce that it would provide new entrant UMTS licensees with the right to roam over existing GSM networks, which had the effect of reducing the value of the licences for new entrants.

Twelve potential bidders entered the fray. These included the four incumbents – T-Mobile, E-Plus (KPN Mobile 77.49 per cent, BellSouth 22.51 per cent) and Viag Interkom (BT 45 per cent, E.ON 45 per cent), with the then Vodafone AirTouch bidding via its new subsidiary Mannesmann Mobilfunk; Group 3G (Marabü Vermögensverwaltung) comprising Orange, Sonera and Telefónica; Talkline, backed by Tele Danmark; Vivendi; and Nets, a wireless software provider with unknown backers. Resellers Debitel – majority owned by Swisscom – and MobilCom intended to bid collectively, but the sale of a 28.5 per cent stake to France Télécom in March 2000 for €3.74 billion meant that MobilCom made its bid in partnership with the French carrier as MobilCom Multimedia while Debitel bid separately. MCI WorldCom and Auditorium Investments, backed by Hutchison Whampoa, also registered to bid.

Shortly afterwards, Vivendi appeared to withdraw but was nevertheless included in the shortlist of eleven which consisted of all of the above bar Nets. Vivendi confirmed its withdrawal and Talkline also withdrew in June. The issue of France Télécom's involvement in two bids as a result of the purchase of Orange, which was against the rules, needed to be resolved, as did the fact that the now renamed Vodafone Group was bidding via subsidiary Mannesmann Mobilfunk but had acquired a stake in France Télécom via the sale of Orange. Orange subsequently withdrew from Group 3G, with its interest split equally among the other co-bidders (leaving Telefónica with 57.2 per cent and Sonera with 42.8 per cent), and joined its parent. Meanwhile, the Vodafone stake was put in trust for the duration of the bidding.

At the end of June, MCI WorldCom decided to withdraw its independent bid and re-enter the auction by joining up with Debitel (and, possibly, to cement the relationship by taking a stake in Swisscom). Auditorium Investments also withdrew with Hutchison planning to re-enter the bidding via a new entity E-Plus Hutchison (see below). This meant that there were only seven bidders left.

Twenty rounds took place during the first two days, but bidding was sluggish and the total offered only $1.6 billion. In a move to prevent bidders from penetrating each other's strategies, only the highest bids in each round were published. After five days and 66 rounds the total was still a meagre $5.7 billion. The fact that Mannesmann, T-Mobile and E-Plus Hutchison (but not Viag Interkom) were intent upon acquiring three blocks – in the case of the first-named pair because they had been awarded less 2G spectrum than subsequent licensees – indicated that no more than five licences would be acquired.

At the end of day seven over $10 billion had been bid, with MobilCom Multimedia the first to reach $1 billion for a single licence. At round 127 (day 10), Debitel withdrew with $28 billion in total on the table and the remaining bidders showing no signs of flagging. The following day, the amount bid exceeded that raised in the UK and the regulator reduced the minimum increase for new bids from 10 to 5 per cent. Both Viag Interkom and Group 3G ceased to bid for three blocks at roughly this point. At the end of day 12 and 150 rounds, E-Plus Hutchison also announced that it was no longer seeking three blocks because the cost had risen to $3.4 billion apiece.

It had been widely assumed that T-Mobile would try to force Group 3G to withdraw, whereas the reality was that Group 3G's determination to carry on bidding induced T-Mobile to rein in its ambitions to acquire three blocks at round 167. The auction concluded abruptly after 14 days and 173 rounds when, to general surprise, the last-remaining interested bidder, Mannesmann, decided that the cost of three blocks was unsustainable. As the regulator had just previously reduced the minimum bid increment to 2 per cent, all blocks ended up costing almost exactly $3.80 billion. Although this outcome had not been seen in earlier auctions, these did not provide the same amount of spectrum in each licence. All licensees bar Viag Interkom subsequently picked up the additional 1 MHz of spectrum for roughly $50 million apiece.

The total raised was $45.85 billion, comfortably above the total in the UK although on a per head of population (pop) basis the $560 (€619) per pop raised in Germany fell short of the $580 (€654) per pop in the UK. Nevertheless, it was considered more than sufficient to depress the share prices of the licensees. Further surprises came at the end of the auction when Hutchison announced that it would be withdrawing from the E-Plus Hutchison consortium and BT announced that it was acquiring the 45 per cent stake in Viag Interkom held by E.ON (formerly Viag and Veba), raising its overall holding to 90 per cent (and subsequently to 100 per cent). In its turn, Telefónica indicated that its shareholding banks – Banco Bilbao Vizcaya Argentaria and La Caixa – as well as Banco Zaragozano and engineering company Abengoa, would each pay at least 5 per cent of its licence fee. Subsequently Sonera announced its intention to reduce its stake in Group 3G from 42.8 to roughly 22 per cent.

It was argued by legal experts that the government had acted as a company in offering licences for sale and hence had arguably breached European antitrust rules by abusing its dominant position. In September, the EU Competition Commissioner launched an inquiry to determine whether rules on state aid had been breached in relation to the 'beauty contests' because the licences had been awarded so far below their apparent market value, but made no comment on the auctions. In October, MobilCom announced that it would be launching a legal action to recover its licence fee, but subsequently retracted, fearful that it might thereby forfeit its licence. In February 2001, three federal states – Bad Würtemberg, Hesse and Bavaria – requested the Federal Constitutional Court to rule on whether they were entitled to any of the licence receipts.

Shortly after, licence holders began discussions among themselves and with the regulator to determine the extent to which they would be allowed to cooperate in rolling out their infrastructure. A new network was estimated to cost roughly $7 billion. Although the licensees other than T-Mobile expressed an interest in building joint networks, the regulator insisted that no alterations to the wording of the licences would be tolerated and that coverage of the first 50 per cent of the population must be achieved independently in every case – which would require roughly 10 000 base stations. Although the licences permitted co-location of masts, this stipulation was expected to create controversy in heavily-built-up areas, as in the UK (see below). Subsequently there were signs that the regulator might take a more liberal view, and Group 3G decided to try and force the issue by announcing that it proposed to save $1.3 billion and 40 per cent of its operating costs by sharing its network with rivals. It subsequently announced that it had reached agreement with E-Plus to lease GSM/GPRS capacity until 2012 in areas where it had no UMTS coverage, mirroring the agreement previously made between E-Plus and MobilCom which declared itself to be in similar talks with all bar T-Mobile. The regulator's response was to instigate an investigation into what sharing would be permissible within the licensing rules.

In April 2001, Debitel announced that it proposed to operate as a reseller and an MVNO from 2003, and had concluded a deal with Vodafone D2 (formerly Mannesmann and due to trade solely as Vodafone in March 2002). A further agreement was signed with T-Mobile in May and with E-Plus in June. Meanwhile E-Plus announced that it would be introducing a variant of partner DoCoMo's i-mode, provisionally branded as d-mode, within the ensuing twelve-month period.

In early June, the regulator published his ruling on network sharing, stating that his intention was not to rewrite the rules but to reinterpret the licence conditions. In essence, operators would be allowed to share the 'active' elements of networks such as base stations, transmitters and controllers provided the operators remained technically separate from one another and hence unable to 'spy' on one another. However, they would not be permitted to use roaming agreements to reach their 50 per cent coverage targets, nor would they be allowed to pool spectrum via a merger. Rather, a merged entity would be obliged to return one of the licences without compensation or to sell it on to a new entrant.

None of the licensees saw fit to raise immediate objections, and shortly thereafter BT and Deutsche Telekom announced a non-exclusive agreement – made binding in September – whereby their respective mobile subsidiaries would share infrastructure roll-out costs in both Germany and the UK, concentrating upon areas with low population density where separate coverage was unnecessary and hence meeting coverage thresholds fixed in licence agreements. The total savings from jointly building roughly 40 per cent of their networks and roaming over the rest were estimated at $2 billion for BT and $3 billion for Deutsche Telekom over a ten-year period taking Germany and the UK together.

In July, the licensees reached an agreement with the Staedetag (council of cities) whereby they would inform local councils of planned UMTS sites and strive to use existing antennas as efficiently as possible. Group 3G also announced that it had signed an agreement with T-Mobile to provide GSM and GPRS services over the latter's network, subsequently moving on to the provision of 3G services. However, heavily indebted Sonera, still in possession of its 42.8 per cent stake, announced that it would not be laying out any further cash on the network, nor guaranteeing any further loans, and that the launch of services would not take place until early in 2003. In September, Group 3G and E-Plus announced that they had reached an agreement on network sharing that encompassed sites, aerials, cables, transmitters and network controllers. However, a dispute then broke out between Sonera and Telefónica Móviles with the latter claiming that Sonera had asked to transfer its stake (though not necessarily all of it), which it had declined to do – a claim denied by Sonera although it did reaffirm its desire to reduce its stake.

At the end of September, fears were expressed that MobilCom's share price indicated that the financial markets placed a value of $1.3 billion on its 3G assets whereas they placed a value of zero on those of similarly placed Group 3G. Furthermore, unless it was bought out by stakeholder France Télécom, which was rumoured to be trying to buy out minority shareholders rather than acquire the 43 per cent stake held by founder Gerhard Schmid – alternatively, it has a call option to buy a majority stake in 2003 – or found some other way to finance its existing $6.5 billion debt load, it was alleged that it would be unable to finance its 3G network roll-out. However, MobilCom announced in January 2002 that it would move to meet the initial 25 per cent coverage obligation by the end of 2003 without linking up with other operators, although it would not rush to meet its obligations any earlier than necessary. Nevertheless, it claimed that France Télécom was contractually obliged to provide MobilCom with loans or loan guarantees covering any funding shortfall until the 'commencement' of UMTS operations, which was interpreted to mean that, in July 2002, France Télécom would be called upon to help refinance a $4 billion bank loan. Furthermore, Mr Schmid claimed that he was technically in a position to force France Télécom and/or Orange to buy a 33 per cent stake from him in specified circumstances that included a dispute over strategy, and it was held that this was precisely what he was trying to engineer. Equally, the need to pay for the shares and take the debt onto its balance sheet was precisely what Orange did not want to do, nor did it want to invest on the scale envisaged by Mr Schmid, although it did want to launch under the Orange rather than the MobilCom brand to establish its brand image. Precisely who is obliged to do what to whom, and under what circumstances, is for now somewhat murky.

For its part, Group 3G launched a GSM service under the 'Quam' brand over the E-Plus network in November – although it proved initially to be something of a marketing disaster with only a few hundred subscribers signing up – with 3G services pencilled in to begin in the first quarter of 2003. However, this timetable would proceed only in accordance with Group 3G keeping to the minimum investment required by its contractual obligations under the licence. In November, Debitel announced that it would supply UMTS services at the same time as its capacity suppliers (T-Mobile, Vodafone and E-Plus), and in January 2002 Vodafone announced that an autumn launch was anticipated – the same as MobilCom. However, T-Mobile promptly responded that it was not looking to launch until the second half of 2003 when its quality standards could be met in relation to technology, handset availability and services, and expressed doubts that its competitors would gain any real advantage through an earlier launch. This prompted Viag Interkom to announce that it too would delay its launch to the end of 2003. Subsequently, T-Mobile specified that twenty cities would be linked up by the end of 2002, commencing with Berlin in February, with full national coverage to be achieved in 2010. By February 2002, T-Mobile had secured over 5000 UMTS sites and intended to raise this to 7000 by the year end.

In January 2002, BellSouth exchanged its stake in E-Plus for an equivalent value of shares in KPN Mobile's parent, KPN Telecom, leaving the latter with 22.51 per cent and KPN Mobile with 77.49 per cent. Subsequently, Group 3G estimated that its cumulative loss before ebitda would amount to $1.75 billion by 2006.

Greece

Four 20-year licences were due to be issued via an auction in July 2001.[50] The reserve price was set at $126 million per licence. The rules stipulated that if four licences were issued, the up-front payment would be 40 per cent of the amount bid, with the rest divided into four annual instalments; if three were issued, it would be 70 per cent up-front; and if two were issued it would be 100 per cent up-front. The main coverage stipulations were at least 25 per cent of the population by the end of 2003 and 50 per cent by the end of 2006. It was intended that a fourth 2G licence would also be issued, and that additional spectrum would be made available to the incumbents via a secondary auction. The 2G incumbents at the time were CosmOTE (OTE 59 per cent, Telenor 18 per cent), Panafon (Vodafone 52.8 per cent) and stet Hellas (Telecom Italia 58.1 per cent), and all applied to participate as did Greek IT group Infoquest, holder of a fixed-wireless licence, although it was initially unclear whether it would restrict itself to a bid for 2G spectrum. All the companies were cleared to bid, but in practice only the incumbents actually did so in respect of 3G licences, with each offering a fraction above the minimum fee. Each then bid for additional 3G spectrum. he final results were Panafon – 20 MHz paired FDD plus 5 MHz TDD for $157 million; CosmOTE – 15 MHz paired FDD plus 5 MHz TDD for $138.8 million; and stet Hellas – 10 MHz paired FDD plus 5 MHz TDD for $126 million. Infoquest subsequently obtained a 10 MHz paired PCS licence, and Panafon and stet Hellas bought 15 MHz paired PCS spectrum plus 5 MHz paired GSM spectrum and 5 MHz paired PCS spectrum respectively. Panafon was rebranded as Vodafone in January 2002.

Ireland

In July 2000, the regulator announced that four 15-year licences would be awarded in April or May 2001 via a 'beauty contest'. The bids would be assessed according to coverage, speed of roll-out, performance guarantees, fees and the promotion of competition. By the beginning of 2001 the market had altered significantly from a year earlier. Eircom subsidiary Eircell was now in the hands of Vodafone which had made Eircom sign an agreement that it would not apply for a licence in competition with Eircell, and Esat Digifone was controlled by what subsequently became the independent mobile former subsidiary of BT,

mmO_2. The third 2G incumbent, Meteor, controlled by Western Wireless International, had only recently begun to provide a service. A dispute then broke out over the price to be set as a minimum licence fee, seriously delaying the award of licences.

In November 2001, the government announced that new regulations would be introduced that would make it possible to install cell sites on public, commercial and industrial buildings without the need for planning permission. This came as a surprise given the trend elsewhere to tighten controls. After much prevarication, it was announced in December 2001 that a 'beauty contest' for four 20-year licences would now take place in 2002, commencing at the end of March with licences awarded in June. Services would be launched in January 2004. The single Class A licence would require 80 per cent population coverage by the end of 2007, while the three Class B licences would require coverage of the five main cities – equivalent to roughly 53 per cent population coverage – by the end of June 2008. Unusually for Europe, the Class A licensee would be required to open up its network to MVNOs, and for this purpose would be awarded additional spectrum. All successful existing 2G licensees would be obliged to offer roaming facilities to new entrants once minimum roll-out conditions were met.

An innovative payment system would be employed, involving an up-front payment followed by a moratorium of three to five years and with deferred payments falling due during the fourth to fifteenth years of the licence. The total cost of the Class A licence would be roughly $45 million, with $11.5 million up-front, while the three Class B licences would cost roughly $34 million apiece with $13.5 million up-front.

Isle of Man

Interestingly, the first network pencilled in to be launched anywhere in the world was that to be provided by BT-owned Manx Telecom for the 75 000 residents in the British crown dependency of the Isle of Man, although few handsets were likely to be available at the specified time of May 2001. The 15-year licence was free and contained no coverage specifications. Rather than roll out a completely new network, the decision was taken to build on top of the existing 23 GSM base stations to which was added a further seven to achieve blanket coverage of the island. It was unclear whether this strategy would create problems of interference, and in any event blanket coverage could not be achieved in this way in much larger countries. However, the kinds of technical problems which were also met in Japan (see below) forced BT to announce a three-month delay in May. In particular, the NEC-supplied handsets tended to cut out as users moved from one cell to another, and NEC – building the network with Siemens via a joint venture called Mobispher – admitted that software glitches were unlikely to be fully ironed out even by the autumn. Furthermore, BT stated that it would commence the service without an 'always-on' Internet connection while operating at only 64 Kbps. The network was readied for launch in October 2001, but with the divestment of mobile subsidiary mmO_2 under way, further delays ensued and the launch eventually took place in early December, using 24 transmitters to cover 85 per cent of the population, but with only 200 single-mode handsets in circulation providing limited services. Charges would not be levied until April 2002 and the full commercial launch would not take place until the latter part of that year.

Italy

The Ministry announced that it would award 15-year UMTS licences to five operators in late 2000, with services to commence in 2002. Each licence would comprise 10 MHz paired plus 5 MHz unpaired. The intention was to conduct an initial 'beauty contest' to determine a short list of bidders, followed by an auction with a minimum fee set for a licence at $1.95 billion (compared to less than $50 million in Germany) to be paid over three years. The 'beauty contest' would be based on capacity and technical infrastructure; services to be provided; commercial objectives including network investment; geographical distribution; and staffing. In order to favour new entrants, these would be given roaming rights on existing 2G networks. MVNOs would be allowed full and fair access to 3G networks once licence winners had recouped their investments.

Licensees would be obliged to provide a service in regional capitals within 30 months from 1 January 2002, and in all major cities within a further 30 months. Once the initial licences were issued, successful new entrants would be issued with an additional 5 MHz paired spectrum for a fixed fee of $683 million. This would bring the total amount of spectrum licensed to 60 MHz paired.

The auction rules included a requirement that offers had to be raised by 5 per cent above the lowest bid on the table during the first ten rounds, and by at least 2 per cent thereafter. However, bids could not be raised by more than 50 per cent above the lowest of the five current bids.[51] The four incumbents – Telecom Italia Mobile, Omnitel (Vodafone 76.9 per cent, Verizon Communications 23.1 per cent), Wind (Enel 56.6 per cent, France Télécom 43.4 per cent) and Blu (Autostrade Telecomunicazioni 32 per cent, BT 20 per cent – now 29 per cent), seemed fairly certain to be awarded licences, and Tiscali, the part-floated

ISP majority owned by Renato Soru, was very anxious to secure the other one via the Andala consortium, so there was unlikely to be room for other new entrants. Initially, three other groups announced their intention to bid, namely Atlanet (Acea/Telefónica/Sonera/Fiat), TU Tlc Utilities (Atitalia/E-tech/ESVES) and Dix.it (e.biscom/Pirelli/Ifil/Banca di Roma/AEM/ePlanet/Securfil). However, Banca di Roma and ePlanet deserted Dix.it to join the IPSE Group, a new bidder headed by Atlanet. In its finalised structure, IPSE 2000 consisted of Telefónica 39.25 per cent, Sonera 19 per cent, Atlanet (Telefónica/Ifil/Fiat/Acea – 12 per cent), Banca di Roma 10 per cent, Xera 5.5 per cent, GoldenEgg 4.8 per cent, Edison 3 per cent, Falck 2 per cent, ePlanet 0.4 per cent and various small companies. Meanwhile e.biscom, Pirelli and AEM dropped out completely, effectively killing off Dix.it as a bidder.

Both Deutsche Telekom and Hutchison were keen to take a major stake in Andala, with Hutchison winning out with a 51 per cent stake and Deutsche Telekom withdrawing from the auction. The rest of the now renamed Andala Hutchison consisted of Tiscali 25.5 per cent/Cirtel 15 per cent/San Paolo-IMI 5 per cent/HdP 1 per cent/Gemina 0.5 per cent/ and Franco Bernabe (BMI) 0.2 per cent. Among the now seven remaining applicants only Tu Mobile lacked a major operator as a backer, but a late registration was logged by Anthill, led by International Last Mile. All bar Anthill were approved as bidders, but Tu Mobile failed to come up with the requisite deposit.

The stakeholders in Blu appeared to fall out over the issue as to whether BT had promised to raise its initial stake of 20 to 51 per cent. It declared itself willing to make only a modest increase, whereupon other stakeholders threatened to withdraw from the auction. The government responded by announcing that such a move might trigger a reduction in the number of licences on offer to four. In the event, Blu entered the auction.

The auction commenced in mid-October. Bidding in the initial rounds was lethargic, but after 11 rounds the total stood at $12.2 billion – equivalent to roughly one third that raised in the UK on a per capita basis. At this point Blu withdrew as stakeholders (including BT) representing less than the 80 per cent of the shares required by the consortium rules were willing to continue. This terminated the auction with the licences awarded to the five remaining bidders, all of which had bid almost identical sums. IPSE and Andala Hutchison became eligible for the additional spectrum. The government threatened to lay claim to Blu's near $2 billion deposit, but Blu argued that it had not contravened the auction rules and had legitimately withdrawn, and hence that the deposit could not be taken as a forfeit. The matter was put in the hands of the courts, with Blu counterclaiming damages on the grounds that the government's action had severely damaged its reputation. In February 2001, the TAR administrative court made its definitive ruling that Blu's deposit should not be forfeit. This was confirmed by the Council of State court in July.

The antitrust authorities also swung into action, initiating an investigation into alleged collaboration between the bidding consortia. This complemented an investigation initiated by Rome magistrates into possible market rigging and 'irregular' behaviour. No evidence of malpractice was uncovered by the antitrust authorities.

There has been a lot of activity since the licences were issued. In November 2000, Sonera reduced its stake in IPSE 2000 to 12.55 per cent. It distributed the rest to Telefónica leaving it with 45.6 per cent. In September 2001, IPSE 2000 announced that it had reached a 2G roaming agreement with Omnitel Vodafone. Subsequently, there were disagreements over the business plan and Telefónica offered to buy out the minority stakeholders for between 20 and 30 per cent of their original outlays. The ongoing bickering among the stakeholders was such as to bring into question IPSE's intention to proceed, and, in January 2002, the decision was made to do the absolute minimum required by the licence conditions and lay off a large part of the workforce. This meant that IPSE would skip the intermediate GPRS phase.

In February 2001, Hutchison Whampoa increased its stake in Andala to 78.3 per cent by buying 25.2 of the 25.5 per cent stake held by Tiscali (which retained the right to repurchase all or part of it before December 2002 but declined to do so in July 2001) and 2.1 per cent of the 15 per cent stake held by Cirtel, for an undisclosed price. Andala was renamed H3G to match Hutchison's brand name used elsewhere, while Omnitel was renamed Omnitel Vodafone. Both confirmed the intended launch of services in the second half of 2002. In August 2001, Hutchison offered to guarantee part of a $4.7 billion bank loan to H3G – in January 2002 $2.9 billion was borrowed from 11 banks and $0.9 billion was provided as vendor finance covering a near-10-year period. Also in January 2002, a failure by CIR to acquire its full stake in H3G left the shareholding as Hutchison Whampoa 88.2 per cent/San Paolo-IMI 5.64 per cent/BMI 2.26 per cent/HdP 1.13 per cent/CIR 1.83 per cent/Gemina 0.56 per cent/Tiscali 0.3 per cent. H3G meanwhile signed a roaming agreement with TIM and also agreed to share its base stations.

The merger of Wind and Infostrada has reduced Orange's stake in the Wind licensee to 26.5 per cent although it has certain rights to extend its stake in the future and is expected to do so. Wind subsequently expressed an interest in acquiring the assets of Blu, including its 1.5 million 2G subscribers, which were put up for sale in early 2002. Wind's competitors for the assets predictably included TIM, H3G and Vodafone, although only H3G appeared to be bidding to acquire the network whereas the others wanted solely the customer base. However, the government promptly banned Wind from making a bid for Blu unless it was just as a temporary holding operation because the government has a stake in Wind's majority shareholder, Enel. The core issue is now whether Blu's GSM spectrum can be sold before it is broken up, should it prove impossible to sell the company as it stands, since a break-up would normally trigger the return of spectrum to the government.

Subsequent upon the fee reduction in France in October 2001, the Italian government refused to follow suit. However, it did open discussions to extend the term of the licences, settling upon 20 years in December 2001. Subsequently TIM announced that it would deploy and launch its network before the end of 2002, but was not expecting much business to be done until 2004.

Liechtenstein

The four 2G incumbents – Viag Europlattform, Tele2/Tango, MobilCom and Telecom FL – were offered the opportunity to roll out a 3G network without a licence fee. Viag Europlattform was the first to accept the offer in February 2000, with formal acceptance in March 2001. Tele2/Tango followed suit in July. MobilCom is considering whether it wishes to accept, with the government as yet uncommitted to the requested change in licence conditions. Telecom FL refused the offer in July.

Luxembourg

Luxembourg intends to use a 'beauty contest' to award four licences – the date is yet to be determined – together with an annual fee amounting to 0.2 per cent of annual turnover.

Malta

2G incumbent Vodafone Malta and recent entrant Mobisle Communications (Maltacom) claim that, under the terms of the Communications Act 2000, they have been awarded licences to operate 3G services. The regulator disagrees, but has yet to explain how it intends to proceed.

Monaco

A licence was awarded to Monaco Telecom in June 2000 without a fee. The total adult population is 27 000. Testing commenced in December

2001 using equipment supplied by NEC and Siemens. However, only 75 single-mode handsets were initially made available.

Netherlands

It was determined that five UMTS licences would be issued in July or August 2000 via an auction that was expected to raise in the region of $9 billion. Libertel (Vodafone 70 per cent) immediately declared its intention to bid but also its unwillingness to pay more than a modest price. Cable operator UPC, MCI WorldCom, Telefónica, Global Crossing and Telecom Italia all withdrew at an early stage. The government then announced that eight companies would be allowed to bid: 2G incumbents Dutchtone (Orange), KPN Mobile (DoCoMo 15 per cent), Libertel and Telfort Mobiel (BT); fixed-wire carrier VersaTel; and newcomers Nogenta Swedish Acquisitions (owned by NTL of the UK), Hutchison 3G Netherlands and 3G-Blue (Deutsche Telekom and the fifth 2G licence-holder Ben Nederland with a 50 per cent plus 1 share stake).

Bidding for the 15-year licences (terminating at the end of 2016), requiring all cities with a population in excess of 25 000 to be connected by 1 January 2007, commenced on 6 July 2000 with the reserve prices for the licences set in total at 470 million guilders ($203 million): the A licence providing 14.6 MHz paired plus 5.4 MHz unpaired; the B licence providing 14.8 MHz paired plus 5 MHz unpaired; and the C, D and E licences providing 10 MHz paired plus 5 MHz unpaired. Unfortunately, two bidders refused to bid. NTL, 22 per cent owned by France Télécom, felt itself to be in an impossible position because France Télécom also had a majority stake in Dutchtone and did not want to bid against itself. NTL was also concerned that consolidation was occurring among bidders which would have prevented it from becoming a force in the 3G sector. For its part, Hutchison Whampoa was in the early throes of forming a joint venture with both KPN and the latter's existing Japanese partner DoCoMo with a view to bidding for future licences via the new entity, and, in the Netherlands, via the existing KPN Mobile bid.

This effectively left VersaTel as the only bidder not already involved in a 2G licence which meant that it would have to incur much higher costs to roll out a network. It responded by taking the government to court to force a review of the auction process on the grounds that it effectively excluded new entrants. Meanwhile, the share prices of the other bidders rose sharply as it was assumed that they would be able to get away with very low bids. This appeared to be supported by the opening bids which exceeded the reserve prices in only one case. By the end of the fourth day the position was somewhat curious with the A licence under offer by Libertel for $198 million, the B licence by KPN for $198 million, the C licence by Dutchtone for $133 000, the D licence by 3G-Blue for $85 000 and the E licence by VersaTel for $171 000. Needless to say, KPN Mobile and Libertel had 2G networks far larger than the other 2G licensees.

At the end of day nine, total bids had reached only $727 million although all of the reserve prices had been exceeded. The auction petered out on day 13 after 305 rounds of bidding – nearly six times as many rounds per day as in the UK – when VersaTel withdrew claiming that it had been bullied by Telfort. Total proceeds came to $2.5 billion, well below early estimates. It was claimed, however, that on the basis of dollars bid per head of population – the larger licences had gone for $42 per pop and the smaller licences for $25 per pop – prices were in line with general expectations before the UK auction. What is now Libertel-Vodafone won the A licence, KPN Mobile the B, Dutchtone the C, Telfort Mobiel the D and 3G-Blue the E.

In the aftermath of the licence awards, the ownership structure of Ben, contributed to 3G-Blue by its owners Tele Danmark and Belgacom while Deutsche Telekom contributed the licence fee and roll-out finance, was altered. Tele Danmark was left with 14.7 per cent, Belgacom with 35.3 per cent and Deutsche Telekom was given 50 per cent minus one share with the option to buy two additional shares at the end of a two-year period to give it majority control. 3G-Blue intends to trade as Ben.

VersaTel complained to the government that the allocation was effectively a 'closed shop'. The government rejected the complaint, but, in October, a Rotterdam court ordered the government to review the auction procedure. In particular, it was alleged by VersaTel that it had been bullied into dropping out and that Telfort had improperly discussed bidding tactics with it – which Telfort promptly denied. At the beginning of November, the antitrust authorities raided the offices of Telfort and VersaTel, while in mid-December the Netherlands parliament advertised for independent investigators to report on the auction. However, in February 2001, VersaTel and BT were cleared of any wrongdoing by the antitrust authorities.

Subsequently, the government declined to offer any financial support to licensees, and the regulator expressed misgivings about network sharing but later announced that this would be acceptable provided competition was maintained. It was reported in October 2001 that KPN Mobile was negotiating a network sharing arrangement with Telfort Mobiel, and this was confirmed in November. In December, Dutchtone and 3G-Blue also announced a network sharing agreement relating to the 'non-intelligent' parts of their networks such as antennas.

Norway

The proposal was to licence UMTS on a national basis via a 'beauty contest' combined with a licence fee of NKr100 million ($11.2 million) upfront and NKr20 million a year for five years. Four 12-year licences were to be issued, each providing 15 MHz paired plus 5 MHz unpaired, with a requirement that 90 per cent of the population in twelve densely populated areas had to be covered within five years of the date of the award of licences. Five trial licences were first issued, although licensees were not guaranteed priority in the 'beauty contest' intended to lead to a launch in 2002 – not imposed by EU regulations in this case as Norway was not a member.

Enitel, the second-largest fixed-wire operator, set out to seek a partner for its bid and formed a 50/50 consortium with Sonera called Broadband Mobile. France Télécom chose to bid via the Orange Norge consortium together with Bredbandsfabrikken, the Norwegian Investor Group and media group Schibsted. Tele1 Europe formed a consortium called BusinessNet with Sweden's Rix Telecom and Western Wireless of the USA. The surprise bidder was the consortium consisting of Orkla, Hakon Gruppen, NorgesGruppen, the Dagbladet newspaper, Hafslund power company, OBOS/NBBL and the postal service (Posten Norge). These four, together with Telenor, NetCom GSM (a Telia-led consortium) and Tele2 Norge (owned by Tele2) constituted the seven registered bidders. Deutsche Telekom refused to bid on the grounds that the market was too small to justify the investment. The winners were Broadband Mobile, NetCom GSM, Telenor and Tele2 Norge.[52] Provision will be made for MVNOs to supply UMTS services.

In March 2001, Enitel announced that it wanted to cut its stake in Broadband Mobile to 35 per cent, but by June had decided to dispose of it in its entirety in order to repay debts. Since cash-strapped Sonera, trading at €6 compared to a high for the year of €97, was not in a position to buy out its partner, it announced that it would abandon the licence which, fortunately, had cost very little to acquire. Once in bankruptcy, Broadband Mobile attracted the interest of Hi3G Access, Orange and Bane Tele, but exclusive negotiating rights were awarded to an unspecified consortium led by local businessman Nadir Naibant. Existing licensees were forbidden from making offers by the regulator. In any event, the purchaser of Broadband Mobile was not guaranteed to be awarded the licence unless it met the original criteria. In mid-September the government announced that it would reissue the licence, but declined to specify either the time, method or price.

On 1 December 2001, Telenor 'technically' launched its network as required under the terms of the licence. However, the network was available only for testing, covered only Central Oslo and conveyed only voice. In February, the regulator expressed reasonable satisfaction with the progress made by Telenor and NetCom GSM, although neither had fulfilled its obligations fully, but requested that Tele2 be fined by the government on a daily basis as it was showing no interest in developing its network.

Portugal

Four 15-year licences were awarded in mid-December 2000. A 'beauty contest' was used to select licensees each of whom was to be awarded 15 MHz paired plus 5 MHz unpaired. The criteria included technical qualifications and financing plans, as well as contribution to the information society, to Portugal's economic development and to sectoral competition. The licensees, who are paying only a nominal fixed licence fee of roughly $90 million (€100 million) plus small annual fees based upon customer numbers, must, however, also invest $768 million in their networks by the beginning of 2002. They are expected to provide roaming between 3G and 2G networks. Coverage is specified as 20 per cent of the population by the end of 2002 and 60 per cent by the end of 2006. As well as the three 2G incumbents – TMN (a subsidiary of Portugal Telecom), Telecel (majority owned by Vodafone) and Optimus (Sonae 45 per cent, EdP 25 per cent and Orange 20 per cent) – the ONI-Way consortium (Electricidade de Portugal subsidiary ONI (in which Banco Commercial Português had taken a 27.5 per cent stake) 55 per cent, Telenor 20 per cent, Grapes Communications and Iberdrola), was expected to win the fourth licence. The other potential bidders were MobiJazz (mainly Jazztel 39.1 per cent, Sonera 31.7 per cent and Mota/Engil 12.2 per cent) and consortia led by Vivendi and Portuguese telco Maxitel.

The ICP was asked to evaluate the respective proposals in mid-November, and ranked the top four in order of merit as Telecel, TMN, ONI-Way and Optimus. They were subsequently awarded the licences. The success of the ONI-Way consortium was partly attributable to the punishing schedule it laid down for itself – 90 per cent population coverage by the end of November 2002. However, Ericsson, contracted to supply handsets to TMN, Optimus and Telecel – now rebranded as Vodafone – announced in July 2001 that these would not be available for testing until the end of 2001 at best with volume production available only in the second half of the year. In mid-October, the government put back the date for the launch of services by one year to the end of December 2002, citing the anticipated lack of equipment prior to that date.

In November 2001, the operators requested permission to share networks in areas with relatively few potential subscribers, including the Azores where TMN, Vodafone and Optimus already shared 2G towers and facilities. TMN also signed a roaming agreement with ONI-Way. Orange is negotiating the purchase of the 25 per cent stake in Optimus held by EdP.

Spain

The regulator awarded four 20-year licences, each of 15 MHz paired plus 5 MHz unpaired via a 'beauty contest' in which bidders were evaluated in relation to their proposed targets for coverage, their available funds and their overall experience and expertise. Six bids were registered by 2G incumbents Telefónica Móviles, Amena Retevisión Móvil (Endesa 28 per cent, Telecom Italia 27 per cent, Unión Fenosa 17 per cent) and Airtel (Vodafone 91.6 per cent, Acciona 5.4 per cent, Torreal 3 per cent), as well as Xfera (Vivendi 31 per cent, Actividades de Construcción y Servicios, Mercapital, Sonera, Acesa, Orange, ACF), Jazztel/Deutsche Telekom and Uni2 (France Télécom). The licences were awarded to the three GSM incumbents as expected and the fourth to Xfera, and only $450 million was raised by way of up-front licence fees although licensees were also required to pay an annual levy of $5 million to the government. The licensees were expected to launch services in 23 cities by 1 August 2001, and a 92 per cent coverage was anticipated by 2005. New entrants were permitted to roam over existing 2G networks, and Xfera signed a five-year GSM and GPRS agreement with Airtel in August 2001. The purchase of Orange by France Télécom triggered an existing agreement that Orange's 7 per cent stake in Xfera would have to be sold on to the other shareholders.

The government was heavily criticised for raising so little money compared to the UK and Germany, and the possibility of increasing the annual fee was widely supported. The government initially saw no merit in the proposal, although it was agreeable to the idea of issuing additional licences, but subsequently announced that it would be applying a new annual levy on 2G and 3G operators in 2001, with the latter each contributing $135 million – a 30-fold increase. The government was technically permitted to impose taxes of this kind, and although the licensees complained vociferously that the fee rise would prevent them from meeting the roll-out timetable, they all later announced firm plans for meeting the roll-out deadlines. Nevertheless, the licensees were widely expected to delay making further investments in their networks, and it was openly admitted at the beginning of 2001 that no services would become available for at least one year.

In early November 2000, the government announced that it would not be issuing any further 3G licences as there was insufficient spectrum available. It would, however, consider the issue of additional 2G licences – providing a particular opportunity for Xfera – and/or the issue of licences to MVNOs. In the event, in April 2001, it decided not to issue new licences but to licence Xfera as an MVNO. At the same time it altered a licence condition – the first time it had happened in the EU – to permit licensees to delay launching their services until 1 June 2002, and announced that it would introduce legislation permitting licensees to utilise spectrum set aside for 3G for improving 2G services.

The licensing process has generated a series of investigations. In September 2000, the European Commission launched an investigation to determine whether the low licence fees constituted a form of state aid. Ferrovial, a member of the unsuccessful Movi2 consortium, was subsequently successful in seeking a High Court review of the licensing procedure. In turn, Telefónica, Airtel (rebranded as Vodafone in October 2001) and Amena filed appeals against the increased annual levy in April 2001. The government responded with a promise that the size of the levy would be reduced as part of the next Budget, and duly announced that it would do so by 75 per cent operative from 2002 and would keep it fairly stable thereafter. Meanwhile, the target launch date of 1 August 2001 came and went and was replaced by 1 June 2002.

At the end of September 2001, it was rumoured that all the shareholders in Xfera bar Vivendi were keen to sell their stakes, with France Télécom the favourite as potential purchaser on behalf of its unsuccessful licence applicant Uni2. Xfera shareholders subsequently announced that they had invested $400 million so far, but were unwilling to proceed any further unless the government awarded Xfera a GSM licence so that it could generate some revenue while waiting for 3G to come on stream. Xfera alleged that its roaming agreement with Vodafone did not permit it to make any profit, but the regulator insisted that Xfera must rely upon roaming for the time being so Xfera was effectively put into hibernation, announcing the lay-off of 70 per cent of its staff in November 2001.

The ownership structure of Auna, parent of licensee Amena, changed in December 2001 when Telecom Italia sold its stake leaving Endesa with 29.9 per cent, Banco Santander Central Hispano with 23.5 per cent, Unión Fenosa with 18.7 per cent and ING with 10.4 per cent.

What was alleged to be the world's first roaming call over a 3G network was made between Vodafone's Spanish and Japanese 3G networks using a single-mode handset in December 2001. Vodafone claimed to have coverage in 20 Spanish cities and that it would launch in the second half of 2002 subject to the availability of dual-mode handsets made by Nortel. However, although the government reiterated that it would insist upon the date of 1 June 2002 as the official launch date for services, subject to unspecified sanctions for non-compliance, it was evident that any launch would involve little more than the connection of testers and employees to networks.

Sweden

Licences were offered in December 2000 with a term of 15 years in respect of the network. Four national licences, each providing 15 MHz paired plus 5 MHz unpaired, were issued. There was a nominal non-refundable entry fee of $10 700. The spectrum would become available no later than 1 January 2002, and 99.98 per cent of the population had to be covered by the end of 2003.

The method for allocating licences was originally to be a pure 'beauty contest' rather than an auction, which was held to favour companies with deep pockets. However, in October 2000, the regulator indicated that a hybrid approach, with an auction following on from a 'beauty contest', was now favoured in some quarters, although it was ultimately decided that there would instead be an annual fee equivalent to 0.15 per cent of turnover. The initial criteria for the 'beauty contest' were that sufficient capital must be available; technical plans must demonstrate reliability, access, speech quality and other service guarantees; business plans must be commercially feasible; and applicants must have suitable experience and expertise. If too many bidders met these criteria, the choice would be made according to planned geographic coverage, speed of network roll-out and service availability.

In addition to the 3G licences, no more than two additional 2G licences, comprising 11.6 MHz paired in the 900 MHz band and 16.8 MHz paired in the 1800 MHz band, were offered to all-comers, although GSM incumbents could not apply solely for the 2G spectrum. In addition to incumbents Telia (bidding for 2 + 3G), Tele2 (2 + 3G) and Europolitan (Vodafone 71.1 per cent – 3G only), and recent PCN licensee Telenordia (BT/Telenor – 2 + 3G), Orange led the Orange Sverige consortium which also contain Skanska, Bredbandsbolaget, Schibsted and NTL (3G only).

A further consortium, Tenora Networks, consisted of Ratos, Nomura, broadcaster Teracom and Glocainet (2 + 3G); Hi3G Access consisted of Investor AB (40 per cent) and Hutchison Whampoa (60 per cent) (3G only); Broadwave Consortium consisted of Tele1 Europe, Western Wireless, Rix Telecom, Suomen 2G/3P Group and You Communication (3G only); Reach Out Mobile consisted of Telefónica, Sonera, Sydkraft and Industrikapital (2 + 3G); and Mobility4Sweden consisted of Deutsche Telekom, Swedish network operator Utfors and ABB Energy Ventures (3G only).

In November, it was announced that the task of choosing licence winners was proving so complex because of the unexpectedly large number of applicants that the awards would have to be postponed until mid-December. The award proved to be controversial since for the first time an incumbent – indeed, in this case much the biggest 2G network operator – in the form of Telia failed to win a licence. The winners were Europolitan, Tele2, Hi3G Access and Orange Sverige. Telia and four others were rejected in the first phase, and while Telenordia scored 3940 out of a possible 3977 points, it still lost out because all four winners scored the maximum. The official drawback to the Telia proposal was that whereas the successful bidders proposed to build between 10 000 and 20 000 base stations by 2003, Telia promised only 4500. Furthermore, whereas Hi3G pledged to spend $3.5 billion on its roll out, Telia pledged only roughly $1 billion.

Telia initiated a formal appeal in January 2001, claiming that the authorities had neglected to take into account either aerial heights or their positioning, but this was rejected by an appeals court in June – at which point Telia chose to take no further action. It was alleged that both Telenordia and Reach Out Mobile had also launched appeals. Meanwhile, and irrespective of the outcome of its appeal, Telia agreed in January 2000 to form a 50/50 joint venture with Tele2's parent at the time, NetCom AB, initially to be called Telia NetCom but subsequently rechristened Svenska UMTS-nät AB. This will lease existing infrastructure from both owners, thereby economising on mast installation, and build out new 3G infrastructure which will be used by Telia and Tele2 as MVNOs providing competing services. The deal is subject to ratification by regulators.

In February 2001, the now renamed Europolitan Vodafone – it has been rebranded solely as Vodafone in April 2002 – and Hi3G Access announced that they had agreed in principle to form a common company to build and maintain a network, mainly outside Stockholm, Gothenburg and Malmö. Called 3G Infrastructure Services (3GIS), this would comprise 70 per cent of their common needs for capacity and each would build the remaining 30 per cent separately as permitted under the regulatory guidelines on co-operation. The companies would remain operationally separate everywhere, but would jointly buy backbone capacity from Skanova, the wholesale arm of Telia – as did Svenska UMTS in January 2002. In May, Orange Sverige signed a letter of intent giving it the option to become a third equal partner in the joint venture. In October, Orange bought the stake in Orange Sverige held by Bredbandsbolaget, thereby raising its own to 85 per cent and leaving the other shareholdings as Skanska (10 per cent), NTL (3 per cent) and Schibsted (2 per cent). Skanska subsequently also sold its stake to Orange.

In March, it became clear that the government's plan to link up every home to a 3G network by the end of 2003 was in trouble, primarily because the government had yet to finalise its blueprint for the roll-out of networks and the licensees had yet to buy enough land for installation of masts and base stations of which some 40 000–60 000 will be needed. Nevertheless, the regulator indicated that it would not accept any excuses for delaying network roll-out by licensees.

In October, state-owned utility Vattenfall which, with fellow utility Birka Energi, was building masts and antennas for 3G licensees in addition to providing network maintenance and the supply of electric power, expressed the view that it was unrealistic to expect demand for 3G services to be forthcoming by the end of 2003, and hence that it was inadvisable for Sweden to insist on such an early date for full population coverage. Meanwhile, the licensees carried on regardless, with Hi3G and Orange discussing the possibilities for network sharing and agreeing to build a joint network in January 2002. Hi3G subsequently contracted Netel to build 250 sites suitable for 3G base stations to add to the 300 sites already contracted by 3GIS.

In December, Europolitan Vodafone opened its UMTS network in Karlskrona, thereby meeting the 1 January 2002 deadline, followed by Orange Sverige in Malmö and Tele2 and Hi3G in Stockholm. At the same time Orange Sverige announced that it had signed a framework agreement to rent antenna sites on existing masts and buildings owned by Sydkraft Bredband in southern Sweden.

Switzerland

The four 15-year national UMTS licences were initially to be auctioned in November 2000 following a pre-qualification period. Each licence provided 15 MHz paired for DS-FDD as well as 5 MHz unpaired for TDD. The minimum price was set at roughly $29 million. Licensees are obliged to cover 50 per cent of the population by the end of 2004.

Ten consortia registered for the auction and were approved as bidders, including incumbents Mobile Com (a Swisscom subsidiary), Orange (France Télécom 85 per cent) and diAx (via dSpeed). Other applicants consisted of Cablecom Management (NTL), Sunrise Communications, Hutchison 3G Europe Investments, Team 3G (One.Tel of Australia, Sonera 3G and Telefónica), Telenor Mobile, T-Mobile and Teldotcom. The last-named subsequently withdrew, as did Cablecom Management, T-Mobile, Telenor Mobile and Hutchison 3G, as well as Sonera and One.Tel from Team 3G leaving it wholly-owned by Telefónica.

The auction was designed to prevent excessive bids, with the incremental bids in each round subject to variation by the regulator and bidders allowed to pass on a round.[53] However, with only five bidders left, excessive bidding seemed an unlikely outcome, and in the event the auction had to be postponed when TDC (formerly Tele Danmark) announced that it was not only provisionally acquiring a 40 per cent stake in diAx from SBC plus up to half of the 60 per cent stake held by diAx Holdings, bringing its total stake up to 70 per cent, but also increasing its stake in fixed-wire operator Sunrise to 89.2 per cent with a view to merging the two holdings into TDC Schweiz. Since diAx and Sunrise were bidding separately in the auction, the prospective merger would leave only four bidders for four licences.

With no other parties showing an interest in entering a new auction, the regulator opened negotiations to fix an annual fee at an acceptably high level. It was extremely anxious to avoid running the auction with the remaining four bidders since it was unlikely to raise much above $100 million compared to the $1 billion hoped for by the government. However, Swisscom threatened to sue if the original auction was not resumed, and in the absence of any demonstrable collusive behaviour by bidders, this became the outcome in early December. Swisscom (licence 1), dSpeed (licence 2) and Team 3G (licence 4) paid the minimum price of $29.3 million while Orange paid $32.3 million for licence 3, representing a meagre total outlay of $120 million.

The situation of Telefónica's subsidiary – renamed UMTS Switzerland – is somewhat precarious because the 2G incumbents have all refused to lease it capacity on their networks. Once Telefónica has covered 20 per cent of the population with a 3G network, the incumbents will be obliged to make 2G capacity available, although the price has yet to be fixed.

Once Swisscom Mobile had won a licence, the Vodafone Group finalised its offer to acquire a 25 per cent stake in the incumbent. In July 2001, Sunrise (on behalf of dSpeed) announced that it doubted its ability to have services up and running by the end of 2002, and that in cities such as Zurich, Basle and Geneva there were regulatory problems with the construction of any kind of network. Every mast must be separately authorised in Switzerland. In November, TDC Schweiz, now holding the dSpeed licence, announced that it would be postponing the launch of services until 2003 or even 2004 as equipment was unsatisfactory and demand inadequate. Swisscom immediately followed suit, but subsequently claimed that it had successfully made test calls on a pilot network in Bern, that it was building 250 antennas to meet minimum licence conditions for population coverage and that it would be able to provide a service to 20 per cent of the population at the end of 2002 – provided handsets were available. In February 2002, the regulator authorised limited network sharing.

UK

The government announced that it intended to auction UMTS licences early in 2000, with all-comers free to bid. Five licences, tenable until the end of 2021, would be on offer, three (C, D, E) with the same amount of spectrum (10 MHz paired plus 5 MHz unpaired), one (B) with slightly more (15 MHz paired) and (A) one with even more (15 MHz paired plus 5 MHz unpaired), with services commencing in 2002 and networks covering 80 per cent of the population needing to be completed by 2007. The A licence would be reserved for a new entrant and a significant barrier to the entry of new competitors would first be removed through the granting of permission for new entrants to roam onto the existing second-generation networks once the own networks were partially completed.

Although existing operators would be free to bid, they would first have to accept licence alterations that would permit this roaming and would be required to close down their analogue networks by 2005 in order to avoid excessive ownership of licences by any one operator. In June 1999, One-2-One was granted leave for a judicial review of these licence changes, and the High Court ruled in August that the government had no right to implement them. The government had this decision overturned in the Court of Appeal in October, with BT Cellnet and Vodafone immediately agreeing to comply.

The closing date for UMTS applications was 6 March 2000. Thirteen applications were received. The great majority of companies involved were, perhaps unsurprisingly, network operators. Of the four 2G incumbent bidders, BT Cellnet and Vodafone AirTouch were so short of spectrum that they were effectively obliged to pay whatever it took. The favourite for the newcomer's licence was SpectrumCo, a consortium led by the Virgin Group and including supermarkets, media groups and investment banks.

The process for selecting the winners was somewhat complicated, but was placed on the Internet for all to see. The so-called simultaneous multi-round ascending auction initially involved two rounds of bids per day (although five per day became the norm), and continued until only five bidders were left in the game. Bidders were allowed to refrain from bidding in up to three rounds, and could take two 'recess days' for consultation once the list was whittled down to eight, but could not bid for more than one licence at a time, were initially obliged to raise their bids by more than 5 per cent above the highest bid in the previous round, and could not switch from one licence to another unless outbid on their first choice. Mistakes would result in expensive penalties.

One of the ironies was that France Télécom was bidding via NTL Mobile – it had a 22 per cent stake in cable TV operator NTL – even though it really wanted to buy Orange, knowing that if it succeeded while Orange did not that Orange would be a much less valuable property, yet if it held off completely from bidding it might not win the auction for Orange either and end up without a licence.

The way the auction unravelled came as something of a surprise. In the first place, the process took longer than expected and, secondly, the amounts bid were much higher than forecast – the original forecast total was under $5 billion – partly because the bidders were offered tax relief on their investments, and it was questioned whether there would be any profits given the prices being bid. It took 106 rounds of bidding to reduce the number of bidders to eight. After six weeks and 110 rounds, BT (bidding as BT3G) and Vodafone AirTouch both raised their bids for the B licence above $6 billion while NTL Mobile and Telesystem International Wireless (TIW) 3G battled it out for the A licence and One-2-One and MCI WorldCom for the C licence. A faltering Telefónica led the way for the D licence and Orange for the E licence. The government reduced the minimum bidding increment to 4 per cent in round 107. Subsequently, BT left Vodafone AirTouch sitting on a bid worth nearly $7 billion for the B licence and switched to the much cheaper C licence, only to switch back in round 124 with a bid of almost $8 billion. MCI WorldCom dropped out and NTL Mobile remained favourite for the A licence. After seven weeks and 133 rounds, Telefónica withdrew and bidding increments were reduced to 1.5 per cent. BT switched yet again to lead the E licence bidding.

With six bidders left, the auction became very critical in round 140 since, first, once the next bidder withdrew the others would be left holding whichever licence they happened at the time to have made the highest bid for, and, second, the only bidder able to bid in any round would be the one without a highest bid for a licence. Interestingly, at that time Vodafone AirTouch's bid for the B licence was $2.9 billion higher than the BT bid for the C licence – a huge amount considering the modest additional spectrum involved.

The auction ended after seven weeks and 150 rounds when NTL Mobile withdrew. This left the winners as TIW 3G (licence A for £4.38 ($6.7) billion); Vodafone AirTouch (licence B for £5.96 ($9.1) billion); BT3G (licence C for £4.03 ($6.1) billion); One-2-One (licence D for £4.00 ($6.1) billion); and Orange (licence E for £4.10 ($6.2) billion).

The implications were interesting. First, Orange boosted its value as a bid target. Second, France Télécom needed Orange more than ever before. Third, Deutsche Telekom reinforced its position in the UK market and got in some useful practice for the forthcoming auction in Germany. Fourth, BT made Vodafone AirTouch pay 'top buck' for its extra spectrum. Fifth, TIW was in desperate need of finance. In fact, this latter problem was addressed quickly as Hutchison Whampoa – recently bumped out of the market when selling its stake in Orange but a 5 per cent stakeholder in TIW – promptly arranged to supply the funds in conjunction with Chase Manhattan and HSBC in return for taking over TIW 3G via its vehicle Hutchison 3G UK Holdings. In December, Hutchison Whampoa, holding at this time a 65 per cent stake in Hutchison 3G UK after transferring a 35 per cent interest to KPN and DoCoMo (see below) and with TIW declining to take a stake, announced that it would be spending $4.34 billion to roll out its network, with services commencing in July 2002. It subsequently signed agreements with Vodafone and BT Cellnet to utilise their existing 2G networks.

In a further interesting subsequent development, the restructured Orange – the original company having been acquired by France Télécom and part-floated – arranged to lease capacity for 2.5G and 3G services from unsuccessful bidder NTL, thereby considerably reducing its roll-out costs while simultaneously positioning NTL as a potential MVNO.

During January 2001, responding to fears among the public that radiation from telephone masts could cause brain damage, the government announced plans to require licensees to apply for full planning permission in respect of every new mast – no minor matter given the need to erect up to 18 000 new masts for 3G coverage – and one county council forbade their erection on its property. Hutchison 3G responded by arranging a 25-year lease on space on 4000 transmission sites with Crown Castle UK, with an option on a further 2536 sites. In addition, BT announced that it intended to lease space for micro-base stations in up to 5000 phone boxes to assist with filling small gaps in coverage. All five licensees expressed a willingness to share infrastructure and the government was thought unlikely to raise any objections because of the radiation issue. However, it announced in April 2001 that it would not tolerate any behaviour that would potentially fall foul of EU antitrust rules.

In response, Oftel, the regulator, stated that licence conditions did not prohibit infrastructure sharing as such, and, indeed, encouraged it in respect of radio masts. The key issue would be whether such sharing would benefit consumers and/or the environment, and anything that resulted in less competition in relation to matters such as network coverage or quality would not be tolerated. The arrangements subsequently entered into by BT and Deutsche Telekom were discussed in Germany above. Allegedly pressurised by licensees, the government warned local authorities in August not to impose a ban on masts nor to insist that they had to be built a minimum distance from schools.

At the end of March 2001, the National Audit Office announced that it would be investigating the licensing process with a view to determining whether it had met all of the objectives set by the authorities. Not long after, the government announced that there was absolutely no prospect of any reimbursement of licence fees.

By July, prospects for a timely roll-out of services had diminished. The now renamed Vodafone Group, in particular, expressed reservations, announcing that it would be building only 750 3G base stations during 2001 compared to its previous target of 1200, and that handset availability was likely to prevent the widespread availability of services until 2003. In September, it caused widespread confusion when it added that because the reduction in the number of base stations would cause congestion, its entire European network would initially operate at only 64 Kbps with upgrading to 384 Kbps no earlier than 2005/06. However, it subsequently announced that 64 Kbps would be the minimum speed at the margins of its footprint, and that it would elsewhere provide by the end of 2002 the 144 Kbps minimum specified by the UK regulator and 3GPP as qualifying for the term 3G, rising to 384 Kbps and eventually to 2 Mbps.

In October, Hutchison 3G signed a second agreement, this time with National Grid subsidiary Gridcom, to use up to 1000 of Gridcom's electricity pylons to house its antennas. For its part, BT formally divided itself into the BT Group and the former Cellnet, now renamed mmO_2, issuing separate shares in each to its existing shareholders. In December, mmO_2 announced that an anticipated lack of handsets meant that its 2G network would not be launched until early 2003. It then signed an agreement to permit Hutchison 3G to roam on its 2G network on the basis of which Hutchison 3G continued to claim that it would be able to launch a basic service in metropolitan areas in September 2002.

4
3G Licensing Elsewhere in the World

Table 4.1 below details the licences awarded so far everywhere bar Western Europe. The sections which follow summarise the state of play by region, covering every country where the government has issued a statement about its plans for 3G.

Table 4.1 Third-generation mobile telephony licences outside Western Europe

Country	**No. of licences**	**Method**	**Month awarded**	**Winners**[1]	**Fee paid $ million**[2]
Australia	Various	Auction	Mar 2001	3G Investments	79
				C&W Optus[f]	123
				CKW Wireless	5
				Hutchison Telecoms[b,d,e]	97
				Telstra[b]	150
				Vodafone Pacific[b]	126
Canada	Various	Auction	Feb 2001	Bell Mobility	480
				Rogers AT&T[b]	260
				Telus[c,g]	237
				Thunder Bay	Negligible
				W2N	8
Czech Repub.	3	Tender	Sept 2001	*Not awarded*[4]	*[4]
Czech Repub.	3	Tender	Oct 2001	*Not awarded*[4]	*[4]
Czech Repub.	3	Auction	Nov 2001	EuroTel Praha	94
				RadioMobil	94
				Not awarded[4]	*[4]

Table 4.1 Continued

Country	No. of licences	Method	Month awarded	Winners[1]	Fee paid $ million[2]
Hong Kong	4	Hybrid–Allocated[3]	Sept 2001	HKT-CSL[f]	[Annual × 15 with minimum of 6.4 p.a.].
				Hutchison 3G HK	
				SmarTone[b]	
				Sunday	
Israel	4	Tender	Dec 2001	CellCom Israel	52
				Partner Comm's	52
				PelePhone Comm's	52
				Not awarded	*[5]
Japan	3	Allocated	June 2000	DoCoMo[e]	Nil
				J-Phone[b,e,f]	Nil
				KDDI[d]	Nil
Malaysia	3	Beauty contest	?	*	*[6]
				*	*
				*	*
New Zealand	5	Auction	Jan 2001	Clear Comm's	4.5 + 0.8
				Hautaki	*[7]
				TNZ	7.5
				Telstra Saturn	3.7 + 0.75
				Vodafone Mobile NZ	4.6 + 0.75
Poland	5	Beauty contest –Allocated[3]	Nov 2000	PKT Centertel	600
				Polkomtel	600
				Polska Telefonia Cyf.[h]	600
				Not awarded	*[8]
				Not awarded	*[8]
Singapore	3	Auction –Allocated[3]	Apr 2001	MobileOne Asia[f]	57
				SingTel Mobile	57
				StarHub[f]	57
Slovakia	3	Auction	Jan 2002?	*	
Slovenia	3	Auction	May 2001	*None awarded*	*[9]
	3	Auction	Nov 2001	Mobitel	88
				Others not awarded	*[9]
South Korea	3	Beauty contest +fee	Dec 2000	KT ICOM	1 090
				SKT-IMT	1 090
				Not awarded	*[10]
South Korea	1	Allocated	Aug 2001	LG Telecom et al	890

Table 4.1 Continued

Country	No. of licences	Method	Month awarded	Winners[1]	Fee paid $ million[2]
Taiwan	5	Auction	Feb 2002	Chunghwa Telecom Eastern Broadband Far EasTone Taiwan Cellular Taiwan PCS	291 302 291 294 220
Thailand	1	Allocated	Feb 2000	TOT/CAT	Nil
UAE	1	Allocated	N/a	Etisalat	Nil[11]

Notes

1 Equipment suppliers as listed in Table 3.1, a–j and k = Samsung.
2 Not all licences provided an equal amount of spectrum, and some provided additional spectrum on a different wavelength.
3 The intended method was not used because of a lack of applicants and the licences were allocated.
4 Twice offered at a minimum of $167 million but no offers made at that price. Re-offered at $92 million and two offers made.
5 See Israel below.
6 See Malaysia below.
7 See New Zealand below.
8 See Poland below
9 See Slovenia below.
10 See South Korea below.
11 See UAE below.

Source: Compiled by author from reports in media.

The rest of Europe

Although most activity in awarding 3G licences has taken place within the European Union for the simple reason that the Commission has mandated deadlines for awarding licences and launching networks, there has been a fair amount of activity elsewhere. This can usefully be explored initially in the context of Eastern Europe.

On the whole, there is a good deal of pessimism about 3G prospects in Eastern Europe where in most cases voice is still regarded as the 'killer application'. The predicted amount of data traffic – roughly 40 per cent of total traffic/revenues – needed to make anything beyond 2G economically viable is seen as a long way off in the majority of countries.

Furthermore, the prospects for new entrants are seen as almost non-existent. Interestingly, joining the EU, as many intend to do before 2010, is not thought likely to make much difference to this prognosis.

The first country to launch a tender was *Poland* where 2G penetration stood at only 15 per cent. It was announced that the proposed UMTS licences would be issued towards the end of 2000 because if this deadline was missed then the tender would have to be delayed until 2002 as no licences could legally be issued in 2001 due to elections. One licence was to be set aside for PKT Centertel (Telekomunikacja Polska (TPSA) 66 per cent, France Télécom 34 per cent), at 'market rates' as part of a deal made in July 2000. The format was to be a 'beauty contest' although it was originally intended that bids could exceed the minimum fee which was set at €750 million, half payable by April 2001 and the remainder in instalments by 2010. Preference would be given to consortia that included a European or global partner, which opened the way for Polska Telefonia Cyfrowa (PTC – Elektrim/ Vivendi/Deutsche Telekom), PKT Centertel, Polkomtel (TDC 19.6 per cent, Vodafone 19.6 per cent) and Netia. NG Koleje Telekomunikacja also intended to bid.

Although the number of licences was initially set at five, each providing 20 MHz, the minister later indicated that he would reduce the fee to €650 million ($600 million) and that he would be willing to reduce the number of licences to four, each providing 30 MHz, but only if the licence fee was also raised. However, he was also willing to restrict bids to the minimum fee established, extend the licence life from 15 to 20 years and allow the first half of the licence fee to be paid before the end of 2001 and the rest over a longer period than previously, although he would not remove the previous requirement that all new licensees be permitted freedom to roam over established 2G networks once new entrants had achieved a 30 per cent penetration rate irrespective of whether the owners of 2G networks had obtained 3G licences.

In mid-October, the minister reverted to the original plan to issue five licences, each of 20 MHz, with the residual 20 MHz to be auctioned off among the five licence holders in 2001. Taken in conjunction with the roaming rules, this so annoyed the incumbents that they threatened not to bid. The minister responded by stating that no further licences would be offered after the initial auction, and that licence holders would be able to seek additional spectrum in 2001. The incumbents responded that they should not be expected to open up their 2G networks until at least a 50 per cent penetration rate had been achieved by new 3G entrants, and appeared to be intent on using the courts to delay the auction until beyond the end of 2000 in the expectation that they would get a better deal in 2002.

In late October, Netia announced that it would not be bidding, citing, in particular, uncertainties about arrangements for roaming. A further dispute then arose when the government demanded a $640 million promissory note from every bidder, only to switch, when opposed by the operators, to a demand for a $115 million bank guarantee which would be cashed in either if a bidder refused to bid, failed to pay for its licence or failed to fulfil other unspecified conditions.

In November, a further blow was dealt when Hutchison Whampoa announced that it would not, after all, be forming a bidding consortium with Polsat and Polpager. In contrast, Telefónica expressed considerable interest as did a consortium led by TU Mobile of Italy and local software firm 7bulls.com, and Polkomtel issued new shares to raise funds to finance a bid. However, in the event, only Polkomtel, Centertel and PTC submitted bids. The government's response was to cancel the auction and award the three licences at the base price of $600 million apiece, of which $225 million was to be paid in advance. A further licence is to be auctioned, probably in 2003 when services are expected to come on stream. However, most unusually, licensees have no obligation to roll out their networks at a pre-specified speed.

Meanwhile, in the *Czech Republic*, where 2G penetration, at almost 70 per cent, is unusually high by the standards of Eastern Europe, the telecoms ministry initially recommended that four 20-year licences should be issued, of which three should go for a fixed fee to the 2G incumbents, EuroTel Praha (Český Telecom 51 per cent), RadioMobil (in which Deutsche Telekom, TDC and Telecom Italia have stakes) and the much smaller Český Mobil (controlled by TIW). Each licence would comprise a minimum of 10 MHz paired. The fee would depend upon the sum raised in an auction for the fourth licence. In total, a sum approaching $500 million was expected to be raised in late 2001. However, the precise method to be used for awarding licences remained subject to political infighting and a new proposal, tabled by the regulator in January 2001, suggested the award of three licences to the incumbents for $134 million apiece followed by an auction for a further licence for at least that sum. This was confirmed by the government in February, stipulating September 2001 as the date for the fourth licence to be auctioned. The incumbents were dismayed by the high fee, and no concrete offers were forthcoming before the end of May, at which point the government agreed to implement a proposal made by the regulator to hire an experienced external consultant. The tender for the first three licences was issued at the end of June with a price tag of $167 million apiece, nearly double that viewed as feasible in non-governmental circles, but the government hoped to persuade the incumbents to apply by promising not to issue further licences if they paid the full sum in cash. Indicative bids were expected by early August with licences issued at the end of September. In the event, only EuroTel and RadioMobil made unspecified offers below the minimum set, but despite this were rewarded with a fall in their share prices. The government's immediate reaction was to state that if any of the incumbents did not eventually apply, its licence would subsequently be auctioned. At the end of September it decided to set the auction date at 30 November and meanwhile put between one and three licences on offer at the original price in October. It rapidly became evident that no bids would be forthcoming if the fee was to be paid in a single instalment, so the government offered to accept half up-front and the rest over five years. Despite this, no offers were forthcoming, with all incumbents arguing that the licence fee was three times above its true value.

Whether any new entrants would be attracted in November was unpredictable, but expectations were low. Shortly before the single-round, sealed-bid auction took place the reserve price was lowered to $93 million, with $13 million payable in December and the rest over ten years with interest. The highest bidder was promised a PCN licence. Licensees were required to launch services by November 2005, at least in Prague. In the event, only EuroTel Praha and RadioMobil submitted bids. In January 2002, the government announced that it would re-issue the third licence on the same terms as previously but declined to specify a date.

Hungary, in its turn, has said little so far other than that although it had intended to issue three licences at the end of 2001, it now proposed a one year delay to take account of the generally adverse conditions in the telecoms sector and the fact that Vodafone, the third 2G licensee, had only recently begun the roll-out of its 2G network. *Romania* plans to issue three licences but must first free up the required spectrum. *Latvia* intends for now to issue a single 3G licence in conjunction with a PCN licence to a new entrant via an auction, with a minimum bid set at roughly $2.5 million and an expected outcome of only roughly $32 million. There are two existing 2G operators, Latvijas Mobilais Telefons (LMT) and Baltkom (owned by Tele2), and these will automatically acquire 3G licences on the same terms as the new entrant. The rule-book was due to be published in April 2001, and a different licensing methodology was by no means ruled out. However, the rule book was delayed and the auction was set back to the second half of 2002. *Estonia* looks likely to issue four licences via a 'beauty contest', whereas *Croatia* intends to offer three licences, probably early in 2002, with a fourth licence held back for a further one or two years. Unusually, the relevant frequency bands are already unoccupied. An initial beauty contest will be followed by an auction. There are two current 2G operators; VIP-Net[b] (Mobilkom Austria 30 per cent) and state-owned Cronet. VIP-Net made an experimental call using Ericsson equipment in November 2001.

In *Slovakia*, the government intended to launch a tender by the end of 2001 for a combined GSM/UMTS licence, with award of the licence pencilled in for mid-2002 at a reserve price of $21 million. The existing GSM incumbents, Globtel (France Télécom 64 per cent) and Eurotel (Slovenské Telekomunikacie – controlled by Deutsche Telekom – 51 per cent, Verizon Communications/AT&T 49 per cent) would also have the option to acquire UMTS licences at a price determined by that received for the UMTS element of the combined licence. However, the launch was delayed to January 2002 with the reserve price for both the combined and separate UMTS licences set at $31.6 million. The 2G incumbents would no longer be guaranteed a licence.

Slovenia, a country with only two million inhabitants but a 55 per cent 2G penetration rate, and where the 2G incumbents are state-owned Mobitel[b] (a subsidiary of Telekom Slovenije), Si.Mobil (49 per cent owned directly by Mobilkom) and Western Wireless, began an auction of three national 15-year licences in March 2001, each providing 15 MHz paired plus 5 MHz unpaired and with a reserve price of $110 million. In addition to the price, consideration was to be given to the expected size and speed of the network and service charges. However, only one bid was received in mid-May, from Mobitel, and since the rules required at least two companies to bid, the auction could not proceed. Subsequently, the government opened a new round of applications, to conclude in September, with the licence fee reduced to $92 million. This time around a solitary bidder would be awarded a licence, but with only Mobitel willing to bid, the near certainty of a 3G monopoly was problematic. In the event, the government awarded a single licence to Mobitel in November, but addressed the monopoly issue by promising to re-offer the other licences at an undisclosed price within eighteen months. A second licensee would be expected to use the existing network. Si.Mobil promptly filed a complaint stating that the auction was invalid because there was only one bidder.

Russia is expected to issue licences covering Moscow and St Petersburg early in 2002. It is unlikely to allocate these using an auction and the fees are expected to be in the region of $3 million and $2 million respectively. Bidders are likely to include MTS (Deutsche Telekom/Sistema), Vimpelcom[a] (Telenor 25 per cent) and Sonic Duo (backed by Sonera). The cost of rolling out a network in Moscow alone has been estimated at $500 million, well beyond the resources of domestic operators, so in order to prevent the licences going to foreign operators these will almost certainly be obliged to have a domestic partner. A further difficulty is that there is insufficient free spectrum, so licences may initially have to be offered 'encumbered'. At the end of 2001, MTS conducted trials with Siemens and NEC, which will subsequently involve the research institutes of the Russian Ministry of Communications, to determine the allocation of spectrum. Licences may also be awarded in the seven federal regions at much the same time, perhaps two per region at a cost of $1 million apiece.

Japan

2G operators	Subscribers 31/12/01
DoCoMo	39 900 000
J-Phone	11 760 000
KDDI (au)	15 840 000
Total	67 500 000 + 5 700 000 PHS
Penetration	56.8%

The route to 3G taken by *Japan* has differed somewhat from that in Europe and the USA. Launched in February 1999, DoCoMo's i-mode service (see Chapter 2) currently operates at speeds no faster than GSM although 28.8 Kbps is about to arrive. However, a key feature is that it is continuously hooked up to the Internet, and had acquired 30 million subscribers by the end of February 2002. The latest i-mode handsets are light in weight; slim; have a design that opens to reveal a screen several times larger than normal; have a long battery life; have had full colour displays since December 1999; have built-in cameras and plug-in keyboards for writing longer e-mails; and recognise spoken commands. I-mode is strictly only a staging post on the road to a fully-fledged mobile data and video service (which is partly why those operators introducing GPRS are not keen on yet another intermediate technology en route to 3G). It is particularly popular with those with no PC-based Internet access, and its main single use so far has been to access daily horoscopes. Furthermore, the cost of fixed-wire connections and Internet access is relatively high in Japan, and huge numbers of workers spend long periods commuting via public transport and hence need entertaining. Unlike its European rivals using WAP, i-mode was designed from scratch to contain a browser that understands HTML. Because it was deliberately built around an open platform, roughly 20 000 websites are accessible. By the latter part of 2000 every new mobile phone in Japan was Internet ready, with Java programming added in January 2001.[54]

DoCoMo (NTT 64.1 per cent) was understandably anxious that its choice of 3G technology, W-CDMA, which it proposed to launch in Tokyo in May 2001 with a three-year national roll-out, became established in Europe, and that underpinned its decision to buy a 15 per cent stake in KPN Mobile in May 2000 despite paying, in cash, what was widely regarded as an excessive price. This investment was added to the 19 per cent stake previously taken in Hutchison Whampoa.

Three 3G licences, each of 20 MHz paired, were awarded without a licence fee. In addition to DoCoMo, the J-Phone companies, in which Japan Telecom, Vodafone Group and BT had a stake at the time, won a preliminary licence in June 2000. BT subsequently sold out to Vodafone and, after restructuring in August 2001, Japan Telecom emerged with a 45.05 per cent stake in J-Phone and Vodafone with a 39.67 per cent stake. However, Vodafone also owns a stake in Japan Telecom, and hence holds a majority economic interest in J-Phone which it has recently negotiated to enlarge to 70 per cent. The third licence went to DDI which was merging with KDD and IDO to form KDDI – with its mobile services trading as 'au' – but this was for a cdma2000 network whereas the others intended to utilise W-CDMA, partly because Kyocera, which acquired the handset business of Qualcomm in December 1999 and is one of only three manufacturers (along with Qualcomm and Nokia) capable of commercialising cdma2000, has a 23.8 per cent stake in KDDI. Permanent licences were awarded after testing was complete. In November 2000, KDDI announced that it would reduce its charge per packet – set between ¥0.3 and ¥0.5 at that time – possibly to one hundredth of that level when it launched its 3G service. Interestingly, it set a target of 600 000 subscribers in the Tokyo region whereas DoCoMo was seeking only 100 000 and hence intended to maintain a high price to choke off demand.

In December, DoCoMo announced that its service, branded FOMA (Freedom of Mobile multimedia Access), would attain national coverage by the end of April 2002. A dense pattern of base stations is easy and cheap to install in Japan compared to Europe. The downstream link would initially operate at 384 Kbps and the return link at 64 Kbps – much faster than for existing i-mode services – but the downstream speed for 3G may not be delivered in practice. Voice quality was guaranteed to be almost as good as with a fixed-wire link – a feature that few other operators worldwide have advertised. The au cdma2000 network, trading as EZweb, was expected to operate initially as cdma2000 1xRTT at 144 Kbps downstream, but would not commence even this limited operation until some time in 2002 in order to provide national coverage at a single point in time, subsequently upgrading to 3G speeds, while the J-Phone network's J-Sky service was not expected to launch in Tokyo until June 2002, with coverage of central and western Japan in October and 90 per cent population coverage pencilled in for the end of 2004. J-Phone's own estimate of the costs to be incurred during the first three years of creating a brand new W-CDMA network was calculated at $5.9 billion.

That things were not running altogether according to plan became clear in March 2001 when DoCoMo announced that, although it had signed contracts with eleven manufacturers for the delivery of sixteen 3G handsets, only four – two each from Matsushita and NEC – would be available by May with others being delayed for up to two years. These delays were ascribed to the complexity of the required software – Matsushita had been working on this for seven years – and the fact that software, semiconductors and base stations all had to be tested for interconnectivity and potential malfunctions. The cost per handset was estimated at three to four times that of an i-mode phone. Subsequently, DoCoMo stated that 3G would initially be restricted to business users, would not reach six million subscribers until March 2004 and should be viewed as 'an evolution, not a revolution in service creation'. However, worse was to come with the announcement in April that technical problems would delay a proper launch until October. Meanwhile, an 'introductory' service would commence with 4000 users – mostly DoCoMo employees – paying for traffic but receiving a free handset and free connection. In October, three handsets were available, costing $165 (data-only), $330 (i-mode with built-in antenna) and $500 (i-mode equipped with a video facility and built-in antenna). However, only about 20 000 handsets were initially available, and they only operated within a 30 km radius of Tokyo. Despite this, DoCoMo announced that it would commence its 3G roll-out beyond metropolitan Tokyo in December 2001 with a view to covering 97 per cent of the population by March 2004. In practice, only 27 000 handsets had been sold by the end of 2001, and with the target of 150 000 sales by the end of March 2002 beginning to recede, DoCoMo announced that it would expand the coverage area at the end of January 2002 and a second time to include major cities in additional regions such as Northern Kanto and Kansai in the Spring.

In February, DoCoMo added that it would be issuing its FOMA D2101V handset – the first to be equipped with i-motion and a videophone, requiring a built-in camera on the face of the videophone plus one on the top edge with a zoom for picking up scenery and so on – at the beginning of March, and that FOMA would be available in virtually every major city and 60 per cent of the populated areas in April, with 90 per cent of all populated areas covered by the end of fiscal 2002.

The situation remains ambiguous for now. On the one hand, initial 3G transmission speeds are, at best, 348 Kbps downstream and 64 Kbps upstream, while on the other hand the new Java-enabled i-mode handsets have been a huge success, with 2.5 million subscribers signing up within three months despite high costs and 2G speeds. Overall, few doubt that 3G will become profitable – the issue is not whether, but when. Vodafone intends to achieve this with a single call rate for all international calls, reflecting its pan-European strategy for 2G roaming. Meanwhile, rumours circulated at the end of 2001 that DoCoMo was deliberately withholding the supply of 3G handsets – as evidence it was pointed out that discount electronics stores were refusing to offer discounts on these phones – because it was afraid that there would be connection failures giving the technology a bad name. In early February DoCoMo announced that it had only 43 000 3G subscribers compared to 27 000 a month earlier. Most subscribers were businessmen and hence the ARPU was a substantial $78 a month, 20 per cent more than the ARPU for i-mode. It was remarked that in rushing FOMA to market DoCoMo appeared to have ignored the fundamental lesson of i-mode, namely that services driven by technology rarely succeed.

China

China has decided that its immense internal market justifies the adoption of an independent solution. Rather than be forced to rely exclusively upon imported technology, it has also adopted the TD-SCDMA standard, developed by Siemens and the Chinese Academy of Telecommunications Technology (CATT), which has been accepted by the ITU as meeting IMT-2000 criteria and is being harmonised with TD-CDMA. Nortel Networks and Motorola have also agreed to develop the technology. However, Siemens has stated that the technology is incompatible with cdma2000 although it can probably be rendered compatible with UMTS. The virtue of adopting TD-SCDMA is that there will be no fights over patents and it will be relatively cheap to roll out. During 2000 a forum was set up for TD-SCDMA led by Datang Telecommunications Technology & Industry Group in which the CATT is a major shareholder.

So far, the government has declined to announce procedures for awarding licences, and although the Chinese carriers have all endorsed TD-SCDMA, it remains unclear whether they will be forced to adopt it – the Ministry of Information Industry has indicated that carriers will be able to choose whichever technology they prefer. A definite decision either way is not expected until later in 2002. The operators are content to develop 2.5G for now. China Mobile Communications and its listed subsidiary China Mobile (Hong Kong) prefers W-CDMA for 3G, and if the licences are awarded before the end of 2002, intends to launch a W-CDMA service at the end of 2003. Meanwhile, the other current 2G licensee, China United Communications (China Unicom), which operates both GSM and CDMA 2G networks, appears to be moving towards exclusive reliance upon cdma2000 for 3G. Given that the award of a further 2G licence has been put back beyond the end of 2001, it may eventually be transposed into a 3G licence.

A major field trial of TD-SCDMA was announced in July 2001 and was expected to last for one year commencing October 2001. Provided the technology was at least as good as its competitors, it would be formally approved by the government. The selection of successful contractors was to have been made early in 2002. However, it became apparent that there were problems with chipsets, handset power consumption and handset design, so development delays may yet prove to be the downfall of this technology although Siemens and the CATT claimed that significant progress had been made in February 2002. Meanwhile, the government issued an invitation to Samsung, Motorola, Lucent and four other equipment suppliers to demonstrate trial cdma2000 1xRTT networks in separate regions. In addition, the Evolium joint venture between Alcatel and Fujitsu successfully demonstrated an end-to-end W-CDMA application in Beijing, and, in December 2001, the government selected eleven vendors to take part in technical trials of W-CDMA in early 2002. After a shortlist of vendors is drawn up, field trials will take place on a variety of networks lasting up to one year.

Subsequent upon the announcement of plans to split China Telecom into northern and southern halves, with the former to be merged with China Netcom and Jitong Communications to form China Netcom Jitong, it was suggested that each half would be awarded a TD-SCDMA licence.

Elsewhere in Asia

The first licence in Asia, providing 15 MHz paired, was awarded without a tender in *Thailand* in February 2000. The joint licensees were the state-owned fixed-wire incumbents, the Telephone Organization of Thailand (TOT) and the Communications Authority of Thailand (CAT). The companies plan to merge their telecoms interests to form Thai Telecommunications later in 2002. Subsequent upon the creation of the National Telecom Commission (NTC), three further licences are to be awarded as late as 2003, when an auction may be involved, with services commencing in mid-2005. Meanwhile, Hutchison Whampoa subsidiary CAT Wireless Multimedia intends to upgrade its existing CDMA 800 MHz licence covering Bangkok and 25 provinces to full cdma2000 capability by the end of 2002, thereby beating the licensees to the starting line for the launch of 3G services.

South Korea set out to award three licences via a 'beauty contest' at the end of 2000 with networks launched in 2002 in time for the World Cup. It fixed lower and upper limits of $890 million and $1.16 billion for the licence fee and eventually settled on $1.09 billion. This sizeable fee, half payable up-front and the other half over a 10-year period – which most analysts considered to be so high as to preclude the possibility of profits for more than one bidder – was expected to favour incumbents Korea Telecom (KT), South Korea Telecom (SKT) and LG Telecom, all of which set up extensive consortia. SK-IMT consisted of SKT (holding 48.6 per cent) together with Pohang Iron & Steel (POSCO) and Shinsegi Telecom (which SKT agreed to take over in June and was given regulatory approval in January 2002 subject to a ban on discounted handsets), while KT ICOM consisted mainly of Korea Telecom (46.59 per cent) together with mobile subsidiaries KT Freetel (KTF – 10 per cent) and KT M.com (5 per cent) which were merged together in May 2001 under the KTF name. Korea Telecom was itself subsequently renamed KT Corp. in preparation for its privatisation.

The government initially allowed bidders to choose their own preferred technology, which could either be W-CDMA or cdma2000. However, it later retracted, stating that if the first two licences were awarded for W-CDMA networks, favoured by all the incumbents, the third would need to be for a cdma2000 network. It allegedly retracted its retraction, but the possibility that a cdma2000 bidder would be favoured caused ISP Hanaro Telecom to submit a last-minute entry even though it was generally thought that it would fail to qualify in the 'beauty contest'.

This surmise proved to be correct. The LG Glocom consortium scored 80.88 per cent, but the two W-CDMA licences were duly awarded in mid-December to KT ICOM (81.86 per cent) and SK-IMT (84.02 per cent). In January 2001, Korea Telecom announced that it would be selling off its 13.87 per cent stake in SK Telecom in stages to finance its 3G services. Meanwhile, the ownership of SK-IMT was adjusted to leave SKT with a 56.68 per cent stake, rising to 61.68 per cent with the purchase of further stakes in Shinsegi, and POSCO with 12 per cent. DoCoMo opened negotiations to take a 14.5 per cent stake in SKT and thereby an interest in a Korean 3G licence, for roughly $3 billion, but the parties were unable to agree upon the price. DoCoMo is now expected to take a stake in KT Corp. when it is privatised.

A further auction for a cdma2000 licence was subsequently scheduled to take place in March 2001. LG Telecom, reeling from a 40 per cent reduction in its share price after the first auction closed, refused to participate, and although Hanaro – which had scored only 56.41 per cent, well below the 70 per cent minimum, in December – agreed to try again, it was not expected to qualify unless it bid in partnership with Qualcomm which promptly announced that it would only bid in conjunction with a 'quality' partner. The government tried to salvage the situation by authorising minority partners in the winning W-CDMA consortia to bid for a cdma2000 licence, authorising cdma2000 bidders to choose their preferred frequency and dialling pre-fix and offering soft loans for roll-out in rural areas. It also announced that W-CDMA licensees would be obliged to provide roaming facilities for all 2G licensees including those using cdma2000 handsets. It subsequently announced that it would not allocate the licence to Hanaro if bidding alone, but only to a consortium. Unfortunately, there was very little enthusiasm to proceed among potential Korean members of a Hanaro-led consortium, and the government decided to postpone further action until a viable bidding consortium had been formed.

In March 2001, it was alleged that the government was intent upon awarding the licence to a consortium controlled by a foreign operator, thought to be TIW, although it was unclear whether TIW would pay the original asking price. This allegation was immediately denied by the ministry. Subsequently, loss-making LG Telecom announced that it was once again willing to bid, but only if the licence fee was significantly reduced. Meanwhile, Qualcomm declined to participate in any locally-led bids. For its part, Hanaro declined to be part of a bid in which LG Telecom would be the majority shareholder, but almost 450 other companies, mostly small, applied to join a LG Telecom-led consortium. Eventually, in July, a proposed 'synchronous IMT-2000 grand consortium' was announced, comprising LG Telecom, a foreign partner (probably TIW) which was trying (unsuccessfully in the event) to acquire BT's 21.76 per cent stake in LG Telecom – between them holding a stake of over 50 per cent – plus Hanaro Telecom (10 per cent) and smaller partners. In late July, Powercomm, the telecoms subsidiary of state-owned Korea Electric Power, agreed to participate, and it was claimed that LG Telecom would be allowed to absorb the consortium before bidding rather than be obliged to create a new business like the other licensees. As a concession, the government agreed that the successful licensee would be allowed to reduce its up-front payment to $169 million with the remainder payable over 15 years in instalments of between 1 per cent and 3 per cent of sales. The eventual offer came from LG Telecom (27 per cent), Hanaro Telecom (11.9 per cent), Powercomm (3.5 per cent), Thrunet and over 1000 small firms – but not TIW. The licence was awarded in August. It is anticipated that LG Telecom will merge in due course with Hanaro Telecom and, possibly, Powercomm. It is alleged that BT is negotiating to sell its stake in LG Telecom to the Hyundai Motor Group.

A dispute then arose between SK Telecom and the LG Telecom-led consortium regarding the allocation of the coveted 'B' frequency band. LG Telecom claimed that the government had promised this to the cdma2000 licensee while SK Telecom wanted the band to conduct W-CDMA trials in conjunction with DoCoMo. In the event, SK-IMT was awarded the 1940–1960 MHz 'B' band, KT ICOM the 1960–1980 MHz 'C' band and the LG Telecom-led consortium the 1920–1940 MHz 'A' band.

It is worth noting that these licences are not technically the first to be awarded in South Korea since SK Telecom launched a cdma2000 1xRTT service in October 2000 but, like GPRS which operates at similar speeds, 1xRTT is best treated for the purposes of this discussion as a variant of 2.5G. By January 2002, there were roughly four million subscribers to cdma2000 1xRTT. It should also be noted that, despite its W-CDMA licence, SK Telecom signed a memorandum of understanding with Qualcomm in June 2001 to co-operate in an attempt to launch a cdma2000 1xEV-DO (Evolution-Data Only) network operating at up to 2.4 Mbps – and hence a 3G service – before the onset of the Soccer World Cup in June 2002, and commenced field testing in November 2001. KTF also conducted 1xEV-DO field trials in December 2001.

It was generally believed that such services as might be available for the World Cup were unlikely to be other than rudimentary. However, in late January 2002, SK Telecom launched its cdma2000 1xEV-DO service in Incheon City and announced that it would be extended to Seoul in February and to all cities hosting World Cup football by April. A video handset would be made available to visiting fans offering speeds at up to 2.4 Mbps. It is evident that W-CDMA has in practice been put on the back burner. KTF has announced that KT ICOM's W-CDMA service will not be launched until at least the end of 2002. The government has agreed that services need not become available until late in 2003 because new 2G subscriptions have slowed down – to roughly one million in 2001 – and 2G networks accordingly still have some spare capacity. For its part, SK Telecom does not seem to be eager to launch its W-CDMA service prior to that deadline.

2G penetration in *Hong Kong* stands at roughly 80 per cent but fierce competition means that only one network, HKT-CSL, is currently profitable. In October 2000, the Hong Kong government finally revealed its intentions, specifying pre-qualification followed by an auction for four 15-year licences, each providing 14.8 MHz paired plus 5 MHz unpaired. The auction was initially expected to take its 'reverse' form – that is, with bids based on the cheapest prices to consumers for an optimum range of services. Most unusually, and ultimately perhaps exceptionally, the number of licences was fewer than the number – six – of operational networks: HKT-CSL (Telstra 60 per cent, Pacific Century CyberWorks 40 per cent), Hutchison Telephone (Hutchison Whampoa, NTT DoCoMo 19 per cent), New World Mobility, People's Telephone (China Resources (Holdings)), SmarTone (Sun Hung Kai Properties 28 per cent, BT 20 per cent) and Sunday. However, winners would be allowed to retain their licences even if they subsequently merged with other operators. Fearful of the cost should it prove successful, investors in SmarTone promptly drove its share price down to its lowest level since listing in 1996. Subsequently, SmarTone took issue with several of the licence conditions, especially that requiring licensees to open up between 30 and 50 per cent of their networks to MVNOs since this would permit an MVNO to end up with more capacity than a licensee.

In May 2001, the regulator, responding to the delays in Japan, announced that the auction would be put back to an unspecified date. The altered form of the auction – a so-called 'blind' version on account of the refusal to let bidders know against whom they are bidding – was unpopular with potential bidders who claimed that its primary purpose was to drive up bids to unrealistic levels. The regulator then proposed that the licence fee be based upon the lowest of the four successful bids – now to be made on a so-called 'revenue loyalty' basis determined by the percentage of annual revenues bidders would be willing to pay to the government. The incumbents maintained that this would still lead to much higher bids than would suffice in an open auction to drive the last unsuccessful bidder out of the auction, and that the government had previously claimed that it was not seeking to maximise its income from the licences.

In response, the government subsequently offered to charge licensees only the amount bid by the company whose offer came fifth by value.

The final auction rules and reserve prices were officially published in July with applications to be lodged by mid-September. Most crucially, there would be four licences; a pre-qualification round would take place; at least 30 per cent of capacity would have to be set aside for MVNOs; all licensees would be obliged to bid a minimum annual fee of 5 per cent of revenues – with bids rising thereafter in 1 per cent increments – with a minimum payment of $6.4 million during each of the first five years and rising thereafter; the price to be paid by all licensees would be the fifth-highest bid plus 0.01 per cent; the auction would be 'blind'; over 50 per cent population coverage would be required by the end of 2007; 2G incumbents would be obliged to provide a roaming facility for new entrants; and no further licences would be issued before 2005. SingTel, the foreign operator thought most likely to make a bid, promptly declined to do so, at least by itself. In the event, only four applications were lodged and HKT-CSL, Hutchison 3G HK (Hutchison Whampoa 74.63 per cent, DoCoMo 25.37 per cent), SmarTone 3G and Sunday 3G were awarded licences at the minimum fee. However, the competition via sealed bid to have first choice of a particular package of spectrum among the four on offer produced very varied bids, with Hutchison 3G HK bidding $300 000, SmarTone 3G bidding $174 000, HKT-CSL bidding $36 000 and Sunday 3G bidding $1200. Hutchison 3G HK declared its intention to roll out its network as quickly as possible whereas HKT-CSL chose to wait 'until the market is ready'. Rumours subsequently circulated that there had been collusion among the potential bidders to limit the number of bids and keep down prices, but these were firmly denied by the regulator.

In October, Hutchison Whampoa announced that it had provided a $57.4 million loan to Hutchison 3G HK, with DoCoMo holding back its contribution pending internal procedures for its authorisation. Subsequently, the regulator refused to respond to requests that network sharing be permitted. In January 2002, newly-established Virgin Mobile (Asia), a 50–50 joint venture between the Virgin Group of the UK and Singapore Telecom, announced that it was inviting 3G licensees to take a minimum 30 per cent stake in the local MVNO that it was setting up. At this point there were already three other established MVNOs: China Motion Telecom, Trident Telecom Ventures – both of which had agreed to lease capacity from Sunday Communications – and China Unicom International. Subsequently, Sunday Communications formed a 50–50 joint venture with the local arm of Royal Dutch/Shell called Shell & Sunday Mobile Communications as a prospective MVNO.

At the end of 2000, the government of *Singapore* announced that it would be auctioning four 20-year licences, one providing 15 MHz paired plus 5 MHz unpaired and the others 14.8 MHz paired in the 1920–1980 and 2110–2170 MHz bands and 5 MHz unpaired in the 1900–1920 MHz band. The reserve price was initially fixed at $86 million but subsequently reduced to $57 million – equivalent to $58 per pop – and the fee would be paid up-front in February 2001 – subsequently set back to April. The simultaneous ascending auction, following on from a simple pre-qualification test, would take place even if fewer than four applicants came forward, with the unsold licences re-offered at a later date. Licensees would be obliged to create an island-wide network by the end of 2004. Licensees would be permitted to make their own choice of technology, sell part of their spectrum and transfer the licence subject to regulatory approval.[55] 3G licensees would be entitled to negotiate roaming agreements with 2G licensees, and the regulator – the InfoComm Development Authority of Singapore – would enforce such agreements if necessary at rates of its choosing. 2G licensees offering a roaming provision would be able to bid for additional PCN spectrum. The initial list of applicants included incumbents Singapore Telecommunications Mobile, MobileOne Asia (M1 – Keppel Group/Singapore Press Holdings/C&W/PCCW) and StarHub Mobile (BT/NTT/Singapore Technologies/Singapore Power), as well as Hong Kong-based Sunday Communications (trading as Sunday Holdings Singapore) and One.Tel of Australia. However, only the incumbents duly applied and the auction was cancelled with the remaining applicants being awarded licences at the reserve price.

In October, M1 announced that it expected to launch trial services at the end of 2002, with commercial services available only at the end of 2003, but could shed no further light on its quest for replacement shareholders given the desire of all of its existing shareholders to sell their stakes. Telstra was reported to be interested, but there was no official confirmation. In January 2002, Siemens, NEC and Itochu conducted a successful W-CDMA trial for SingTel.

Taiwan, where 2G penetration stands at an extremely high 90 per cent due to multiple subscriptions, originally intended to licence three national networks in April 2001, probably via an auction even though all previous fixed-wire and mobile licences had been awarded via 'beauty contests'. Potential bidders included Chunghwa Telecom[f] (97.4 per cent state owned), Taiwan Cellular (Pacific Wire & Electric Cable, Verizon Communications 13.5 per cent), KG Telecom[f] (Koo's Group, DoCoMo 21.4 per cent) and Far EasTone (Far Eastern Textile, AT&T Wireless – bidding via subsidiary Yuan-Ze Telecom). The regulator commissioned a report by Nomura which recommended that four 15-year national licences be issued via a multi-round auction, but the transportation and communications ministry opted for five, with the fifth ('E' licence) replacing Chunghwa Telecom's outmoded AMPS licence and, unlike the other four, based upon cdma2000 rather than W-CDMA. The award of licences was pencilled in for January 2002. In mid-December 2001, a total of seven applications were made, with Eastern Broadband Telecom, Taiwan PCS Network (Yulan Motors) and the 3GO consortium led by First International Computer joining the others listed above. Since this was the first time in Asia that there were more applicants than licences available, it became possible to hold an auction with minimum bids set at $120–220 million depending upon the bandwidth on offer. In October 2001, as a preparatory move, KG Telecom ordered a core platform from NEC similar to that used for i-mode in Japan and capable of sending data at 114 Kbps over its GSM/GPRS network, only to withdraw its 3G licence application in December.

The auction commenced in mid-January 2002 and was scheduled to end when none of the six bidders submitted new bids. At the end of day two, only Far EasTone, with the high bid of $131 million for the 'B' licence, faced a challenge – in this case from 3GO which had switched from contesting the 'E' licence. The uncontested bids were by Taiwan Cellular for the 'A' licence, Taiwan PCS for the 'C' licence, Chunghwa for the 'D' licence and Eastern Broadband for the 'E' licence, all at little more than the pre-determined minimum. Subsequently, the auction proceeded slowly on the basis of the smallest permissible incremental bids. However, by day 12 the total amount bid exceeded the minimum by 28 per cent, and concerns began to be expressed about over-payment. Bids nevertheless continued to escalate. The auction ended after 19 days and 180 rounds when 3GO withdrew. In total, bids came to $1.4 billion, almost half as much again as the reserve price. Far EasTone won the 'A' licence for $291 million; Taiwan PCS the 'B' licence for $220 million; Taiwan Cellular the 'C' licence for $294 million; Chunghwa Telecom the 'D' licence for $291 million; and Eastern Broadband the 'E' licence for $302 million.

Malaysia originally intended to offer up to four licences with some element of competitive bidding involved. It seemed unlikely that it would allocate licences before the end of 2001 as the government believed that GPRS was adequate for current purposes. There are five 2G incumbents – state-owned Telekom Malaysia, Maxis Communications (Usagha Tegas 70 per cent), DiGi.Com (Telenor 61 per cent), Celcom (Technology Resources Industries – TRI) and TimedotCom (Time Engineering 55 per cent, government 20 per cent), of which the latter two were considered unlikely to bid by themselves. In early October 2001, the government announced that it no longer favoured an auction and would probably allocate the licences at a low fixed price. It was anticipated that, depending upon the number on offer, an equal number of bidders, some in the form of consortia, would be created. That number was confirmed as three in early November, with selection via a ‘beauty contest’ emphasising the financial strength of applicants and their roll-out plans. Those existing operators forming joint ventures or consortia would be prioritised. The licences would last for 15 years and cost $13 million, payable in instalments. The tender would be issued in February 2002 and licences issued in July, with a launch date of late 2003. However, there was no announcement relating to the preferred technology.

In the event, the government allegedly decided unofficially to award licences to Telekom Malaysia, DiGi.Com and Maxis Communications in December 2001, possibly waiving the licence fee. This was officially denied by the government and formal confirmation of the licence allocation is not expected until June 2002. It was announced in January 2002 that all licences would be for W-CDMA. Although Telekom Malaysia intends to install a cdma2000 network, this will be solely devoted to improve access to rural areas and for fixed-wireless services. In February, the Malaysian Communications and Multimedia Commission released a Marketing Plan for public comment with frequency bands dedicated to 3G assigned as 1920–1965 MHz, 2010–2025 MHz and 2110–2155 MHz. Subsequently, it looked increasingly likely that Telekom Malaysia would take over heavily-indebted TRI, and thereby acquire Celcom, while Time Engineering put TimedotCom up for sale with Maxis the likely buyer. These developments will neatly resolve the issue of how to achieve three consolidated licensees.

A similar procedure is also the intention in *Indonesia* where five licences are pencilled in for allocation via a ‘beauty contest’ in 2002, and in the *Philippines* where at least five licences are to be awarded. A total of 60 MHz is to be made available, in part by re-allocating spectrum previously set aside for 2G. The position in *India* is that spectrum is being cleared in preparation for an auction of between 3 and 5 licences, probably some time in 2002.

The USA

In 1996, a law was passed in the *USA* with the laudable intention of encouraging a shift from analogue to digital TV. Unfortunately, as it transpired, it specified that the 137 TV stations occupying UHF channels

60–69 were not obliged to return their analogue frequencies until the later date of either the end of 2006 or when 85 per cent of their customers had switched over to digital, while collecting, for free, a great deal of new, highly valuable, spectrum.

This effectively meant that most of the relevant spectrum would remain in the hands of a group of technologically backward companies which would have little incentive to hand it back even though it would earn them very little by way of direct revenues or profits. This is because so long as its analogue signal is broadcast, a broadcaster is entitled to a free channel on the local cable network as a consequence of Federal Communications Commission (FCC) rules that require cable subscribers to be able to access all local channels.

The FCC could, in principle, have confiscated the spectrum, but that would have brought it into conflict with owners of the huge numbers of US homes that have neither cable nor digital aerials. Alternatively, the law could have been changed, but that would have brought Congress into collision with the same group not to mention the powerful broadcasting lobby.

As a consequence, most of the optimum spectrum for IMT-2000 in the USA – the 700 MHz band – remains occupied by broadcasters, while 1710–1860 MHz is occupied by the military, 2110–2150 MHz is used by schools and health care centres and 2500–2690 MHz is reserved for Multichannel Multipoint Distribution Services (MMDS) and the Instructional Television Fixed Service. This means that when the FCC comes to auction spectrum specifically designated for 3G services, it will probably be obliged to offer it 'encumbered' in most areas. The winners – assuming anyone wants it in the first place – will then have to persuade (i.e. bribe) the owners to release it, but if they prove unwilling to do so because, in the case of broadcasters this would cause them to lose their cable channel and hence, in effect, go out of business, the spectrum will remain useless for 3G for years to come.

At the end of July 2000, the FCC voted 3–2 to delay the initial 3G spectrum auction – covering 747–763 MHz + 777–792 MHz – until March 2001, even though, under Budget law, the auction was legally required to be held during 2000. In the aftermath of the publication of a Council of Economic Advisors Report entitled *The Economic Impacts of Third Generation Wireless*, which extolled the benefits to be derived from 3G, President Clinton signed a Presidential Memorandum asking for the spectrum issue to be resolved with all due speed[56] and a series of studies were set in hand, ostensibly to identify the relevant spectrum by July 2001 and to auction it by September 2002.

The 'C' and 'F' licence auction

In December 2000 there was an auction for regionally-based spectrum in the 1900 MHz PCS band. Four hundred and twenty-two 'C' and 'F' block licences were on offer – with 90 confiscated from NextWave Telecom for non-payment in an earlier auction but subject to a legal appeal. One hundred and seventy licences were reserved for entrepreneurial companies such as Salmon PCS and Alaska Native Wireless. This auction allowed five of the six major operators – AT&T Wireless[g], Cingular Wireless[b,f,h] (BellSouth/SBC), Sprint PCS[k], Verizon Wireless[c] (Verizon Communications/Vodafone) and VoiceStream Wireless (under offer from, and later acquired by, Deutsche Telekom) to fill in gaps in their networks, and, in conjunction with spectrum swaps, to provide (near) national coverage. The sixth, Nextel Communications, did not operate a PCS network. After 101 rounds of bidding the total bid reached $16.85 billion. Verizon Wireless offered $8.78 billion for 113 licences, including $2 billion apiece for two 10 MHz licences in New York. Alaska Native Wireless, with ties to AT&T Wireless which had itself withdrawn, offered $2.89 billion for 44 licences including $1.5 billion for the third New York licence, while Cingular Wireless (bidding via Salmon PCS) offered $2.35 billion for 79 licences but failed to acquire one in New York where it was short of spectrum. Interestingly, the overall average cost per megahertz per pop was $4.08 compared to an average of $4.40 in the UK and German auctions, although it was much higher in the main US cities – up to $11.30 in New York.

In mid-February, a number of entrepreneurs brought a court case claiming that the auction result should be annulled on the grounds that some allegedly entrepreneurial companies had been unfairly bankrolled by incumbents. Cingular Wireless, for example, had taken an 85 per cent stake in Salmon PCS, paid for its licences and intends to build its network. Only Verizon Wireless, among the five majors, bid exclusively in its own right. The FCC had claimed that ties between entrepreneurial companies and incumbents were not ultra vires, although such ties could not include a controlling interest held by the incumbent. According to the court case brought by TPS Utilicom Services at the end of September 2001, both Alaska Native Wireless and DCC PCS had ties to AT&T that disqualified them from bidding.

It is possible that 3G services will be launched using the 'C' and 'F' block licence spectrum, which is already available for use. However, this is problematic. Most importantly, after a series of setbacks in the courts, NextWave finally obtained a positive judgement from the Court of Appeals for the District of Columbia in June 2001 which ruled that the

FCC had violated the provision of the bankruptcy code prohibiting government bodies from revoking licences solely because a licensee had failed to pay debts dischargeable in bankruptcy. The FCC launched an appeal in the Supreme Court – which was unlikely to be resolved before June 2002, if accepted – on the understanding that if this failed it would be obliged to return the licences to NextWave or pay NextWave compensation out of the auction proceeds. Alternatively, the appeal launched by three operators – including Verizon Wireless, which had acquired 68 ex-NextWave licences and hence had most to lose – on the grounds that NextWave had not been properly qualified to bid for licences in 1998, and was not properly qualified to re-acquire them, had some hopes of success, although not very high. As a precaution Verizon Wireless simultaneously used its partner, Valley Communications, to put in a bid worth $3 billion for part of NextWave's spectrum. However, NextWave secured $5.5 billion of financing, sufficient to pay all of its creditors in full, and petitioned to be released from bankruptcy proceedings. This led the thwarted licensees to table a proposal whereby they would pay $10.1 billion to the FCC and $5.75 billion to NextWave to resolve the issue.

It began to look possible that NextWave would surrender its licences at the beginning of December 2001 in return for a lump-sum settlement, but Verizon Wireless raised objections at the end of September, stating that it would be willing to pay $3.1 billion in December, but not the remaining $5.3 billion of its original bid until the later of either one year after all proceedings were final or 31 December 2002 – thereby allowing time for an IPO of Verizon Wireless. It also asked for the return of $1.5 billion of its $1.7 billion deposit plus interest so far. The possibility of a speedy resolution faded with the Supreme Court appeal launched by the FCC in late October which held that bankruptcy law should not supersede statutes designed to allocate the airwaves 'effectively and efficiently in the public interest'. However, this case was in principle rendered irrelevant by the provisional agreement tabled in late October whereby the $15.85 billion would be paid in instalments, with letters of credit covering the money needed to pay off NextWave issued in January 2001 and the rest made payable in June 2002, whereupon NextWave would receive its money. The deal was agreed to by the FCC in mid-November, and passed on to be ratified by the Justice Department – which ratified it in November – and Congress. However, if the agreement was not passed into law by the end of 2001, thereby eliminating the possibility of further litigation, it would technically fall by the wayside. Unfortunately, that was, in the event, the outcome because there was opposition to handing over such a huge sum to an insolvent start-up, and it remains unclear what will happen in 2002. Congress may or may not ratify the agreement and Verizon Wireless may or may not acquire the badly needed spectrum in some other way. For now, the thwarted operators have asked for the return of their deposits totalling $3.2 billion, and it looks probable that 97 per cent of these will be returned, leaving the bidders with a residual interest in case the FCC appeal was successful. However, Verizon Wireless has gone so far as to open legal proceedings to recover its full deposit, claiming that the delays in proceedings have effectively devalued the licences. Meanwhile, NextWave has proceeded to launch a trial service in 60 markets with the other 35 to follow on by mid-2002.

A further 45 'C' and 'F' band licences were awarded to Salmon PCS in October 2001. Nevertheless, the 'C' and 'F' licence spectrum was generally regarded as inferior to spectrum in the 700 MHz band where licences with potentially nation-wide coverage would at some point be on offer with no caps on the amount of spectrum that could be obtained by any one operator – there was at the time a 45 MHz cap in the PCS band in a major city market and a 55 MHz cap in a rural market. When this auction eventually transpires, it will provide two licences in each of six regions – one set providing 10 MHz paired, thought to be sufficient for W-CDMA, and the other 20 MHz paired. Nevertheless, divided up among the six, the amount of spectrum available in the 1900 MHz band and such part of the 700 MHz band as can be cleared for 3G provision before the end of 2002 is unlikely to prove anywhere near adequate for a full range of competing 3G networks. The FCC accordingly began a review of the spectrum caps knowing that, if these were lifted, there was a distinct possibility that there would be consolidation among the six main competitors in the mobile market. The FCC duly announced in November 2001 that the spectrum cap in urban markets would be lifted to 55 MHz with immediate effect and that all spectrum caps would be eliminated on 1 January 2003.

Stance of operators

The National Telecommunications and Information Administration (NTIA) at the Commerce Department initially took the view that up to 168 MHz of spectrum could eventually be made available in the 698–746, 746–764, 776–794, 1710–1755, 2110–2150 and 2160–2165 MHz bands, although interference from satellite signals would need to be addressed. Initially, the uncertainty that surrounded the issue suggested that there would be better prospects for EDGE in the USA compared to Europe, especially in relation to the 2G TDMA networks of AT&T Wireless and Cingular Wireless, although a complete GPRS overlay would first need to be installed. However, with DoCoMo agreeing to purchase 16 per cent of AT&T Wireless for $9.9 billion at the end of 2000, it came as no great surprise when AT&T announced that it would be building out a GSM overlay on top of its TDMA network. Subsequently combined with GPRS, this would boost the data rate of AT&T's Pocketnet service to 115 Kbps. At some point in 2002–03, the overlaid network would migrate to EDGE or, more probably, to W-CDMA. Cingular Wireless has also decided to follow the same route, supported by its recent agreement to utilise VoiceStream's New York network (see below). In January 2002, rural wireless operator Dobson Communications joined these two, its existing main roaming partners, via new roaming agreements based upon GSM and GPRS. Shortly afterwards, AT&T Wireless announced that it had formed a joint venture with Cingular Wireless, subject to regulatory approvals, to build a GPRS-enabled network along 3000 miles of interstate highways in the Midwest and West, and also that it had formed a roaming relationship based upon GPRS with Rogers Wireless in Canada.

Both Sprint PCS and Verizon Wireless announced that they intended to make initial investments in 3G networks during 2001. Verizon unveiled the award of a $5 billion contract to Lucent Technologies (together with Motorola and Nortel Networks) for a cdma2000 1xRTT network – which came as something of a surprise given that co-owner Vodafone was introducing W-CDMA in Europe where Lucent was being left with little more than crumbs from the 3G table. It appeared that unless majority-owner Verizon Communications agreed to the use of W-CDMA by Verizon Wireless, Vodafone customers would have to switch handsets when crossing the Atlantic as the prospects for a dual-mode handset seemed to be slim despite Vodafone's pleas to handset makers to resolve the problem. However, as noted above, Qualcomm announced in December 2001 that it was about to develop a chipset that would, sometime in 2003, underpin such a handset. Verizon Wireless is now talking about a launch in 2004. Meanwhile, Sprint PCS intends to roll out a cdma2000 1xRTT network in conjunction with Lucent, Motorola, Nortel and Qualcomm, and, together with the other prospective 28 cdma200 1xRTT operators, received its first handsets, made primarily by Kyocera Corp., LG Electronics and Sierra Wireless, in early January 2002 – it subsequently added Samsung, Sanyo and Hitachi to the list. It predicts speeds of 144 Kbps in 2002 and 307 Kbps in early 2003 using the 1xRTT network, followed by 2.4 Mbps in late 2003 using 1xEV-DO (Evolution-Data Only) and 5 Mbps in 2004 using 1xEV-DV

(Data & Voice). It alleged that it would manage with its existing spectrum, thereby minimising costs, but both Verizon Wireless and Sprint PCS run the risk of becoming technologically isolated. Cingular Wireless, in contrast, which needs to convert its GSM and TDMA 2G networks, has refused to declare its hand on the grounds that it is still assessing the alternative technologies on offer. AT&T Wireless seems determined to convert its TDMA network to W-CDMA – although it must first roll out a GSM network and upgrade it via GPRS (which it launched in Seattle in July 2001) before switching to 3G. AT&T Wireless claims that, taking into account spectrum held by its partners, it should be capable of providing GPRS in the top 100 US markets and 3G in the top 70.

Recent developments

In March 2001, subsequent upon his election as President, George Bush proposed that the FCC should have until 30 September 2004 to deposit the proceeds of the auction of channels 60–69 3G spectrum. These, together with the auction of spectrum occupying channels 52–59 in the 698–746 MHz range, were expected to yield $11.1 billion. As noted, broadcasters were not obliged to give up their spectrum on a market-by-market basis before the end of 2006 unless either all four of the largest networks and affiliates were broadcasting in digital, digital-to-analogue converter technology was generally available or fewer than 15 per cent of the households in the market were not receiving digital signals. This proposal, together with a provision to charge TV broadcasters $200 million a year in fees from 2002 to 2006 to hasten the switch to digital signals, needed approval by Congress which, on past experience, seemed unlikely to be forthcoming. It was scarcely surprising, therefore, that the FCC voted for an indefinite extension to the proposed auction in July 2001.

In a further blow to 3G prospects the FCC published its *Final Report. Spectrum Study of the 2500–2690 MHz Band: The Potential for Accommodating Third Generation Mobile Systems* at the end of March 2001. This concluded that the 2500–2690 MHz band was not used consistently throughout the USA but was heavily used in metropolitan areas and hence could not easily be cleared for 3G use. This was followed immediately by *Final Report. The Potential for Accommodating Third Generation Mobile Systems in the 1710–1850 MHz Bands: Federal Operations, Relocation Costs, and Operational Impacts* from the NTIA which concluded that it would be equally difficult to shift defence operations occupying 1710–1860 MHz to other frequencies. This was something of a self-fulfilling prophecy since the US Army and Navy promptly refused to vacate the spectrum under any circumstances, while the Air

Force demanded compensation of $3.2 billion. Interestingly, the likes of AT&T Wireless and Sprint PCS remained fairly unperturbed, arguing that they could anyway manage to cover most of the population using existing spectrum – although their comments preceded the successful NextWave appeal discussed previously.

On a more positive note, the majority of the broadcasters occupying channels 60–69, acting as the Spectrum Clearing Alliance, obtained agreement from the FCC to permit them to negotiate privately with interested parties to establish the open market value of the spectrum and, if they agreed to vacate the spectrum, that they would be able to continue analogue broadcasts until the end of 2005. However, this had to be set against the FCC's announcement at the end of September that it would not be reallocating spectrum in the 2500–2690 MHz band occupied by the fixed-wireless operations of schools and hospitals, WorldCom and Sprint, although it would authorise its partial use for 3G. In October 2001, the government ordered yet another investigation into spectrum use, concentrating upon the 1710–1770 and 2110–2170 MHz bands, with a report-back date of late Spring 2002. For its part, the FCC announced that, after four delays, it would finally be going ahead with the auction of spectrum in the 700 MHz band in June 2002, with twelve licences to be put on offer, even though the issue of how, and when, to remove incumbents had still not been resolved. In December, the FCC started the process for selling off the spectrum covering channels 52–59, but refused to sanction the kind of private arrangements permitted in respect of channels 60–69. Since channel 52–59 broadcasters appeared to be wholly disinclined to vacate their spectrum before 2006 at the earliest, it was unclear exactly what the FCC hoped to achieve.

Meanwhile, Verizon Wireless announced that it had begun trials on its cdma2000 1xRTT 'Express Network' in Philadelphia, and launched the service in January 2002 across 20 per cent of its overall network. The data speed was estimated at between 40 Kbps and 60 Kbps. For their part, VoiceStream Wireless and Cingular Wireless agreed to share their respective networks in New York City, California and Nevada as from early 2002. This was potentially significant because it meant that Cingular Wireless was now, like VoiceStream and AT&T Wireless, taking the GSM route to W-CDMA, but Cingular was quick to point out that it would evolve to EDGE and make do with that for a period of three to five years without committing itself to anything further.

In assessing the prospects for 3G in the USA it is also helpful to bear in mind that mobile phone users are normally charged to receive calls, and hence often leave their handsets switched off. Furthermore, with home computers relatively pervasive, and free local calls on offer, it often makes sense to provide the younger generation with separate phone lines and a subscription to an instant messaging service. It may also be noted that the relative ubiquity of handheld computers means that many Americans tend to think in terms of adding wireless capability to such devices rather than adding Internet connectivity to mobile phones. This also helps to explain the view that is increasingly expressed to the effect that 3G will to a considerable extent be effected in the USA via so-called Wi-Fi (Wireless Fidelity – see Chapter 5). In November 2001, VoiceStream applied to acquire the assets of bankrupt Wi-Fi operator MobileStar Networks to help fill gaps in its coverage, but this is too small an operation to provide clear evidence of the role that Wi-Fi will play.

Elsewhere in North, South and Central America

Meanwhile, in *Canada*, seven companies – Bell Mobility, Microcell Telecommunications, Rogers AT&T Wireless, Telus Mobility, 3050443 Nova Scotia (Sprint), W2N (Wireless2Net) and Thunder Bay Telephone – registered an interest in the 40 MHz of spectrum per market being made available via 64 licences, each of 10 MHz bandwidth, in 16 regional/local markets, using spectrum in the 2.5 GHz band. The list of accepted bidders included all bar W2N which was asked to supply additional information. The auction began in mid-January. It was not anticipated that large sums would be raised given the modest size of the market and the ongoing process of consolidation that had caused some likely bidders to lose their independence. In particular, the acquisition of Clearnet by Telus meant that Telus was obliged to return 20 MHz of spectrum in Alberta, British Columbia and Quebec that exceeded its permitted upper limit of 55 MHz. Furthermore, the spectrum may well be used for 2G and 2.5G purposes, at least initially, so it is unclear when 3G will first come into being.

The first to drop out from the auction was Microcell Telecommunications, and was quickly followed by Sprint. In total, the bids were worth $1 billion: Bell Mobility bid $480 million for licences in Alberta, British Columbia, Newfoundland, Saskatchewan and Southern Ontario where it reached the 55 MHz cap; Rogers AT&T Wireless bid $260 million in total; Telus bid $237 million in total; W2N bid $7.6 million in total; and Thunder Bay bid $400 000 in northern Ontario. On average, the price was only $0.94 per megahertz per pop, but much higher in

Southern Ontario. Eight licences did not attract bids. In November 2001, Telus demonstrated a cdma2000 1xRTT network and announced plans to offer 3G services in early 2002. Subsequently, Rogers Wireless and AT&T Wireless signed new roaming agreements based upon GSM/GPRS and intend to move towards a W-CDMA-based 3G network linking the companies across the US–Canada border.

In South America, the *Brazilian* regulator Anatel has announced plans to auction licences early in 2002 using the 1.9 GHz PCS band. However, only 30 MHz paired has initially been set aside, well below the level in most of Europe. As from 2005, existing 2G licensees in the PCS band will cease to have exclusive use of those frequencies, which could then be reallocated for 3G use. Meanwhile, the successful licensees in the recent 'B', 'D' and 'E' band auctions will be able to switch from 2G to 3G as soon as equipment for the latter becomes available – should they so wish. Portugal Telecom-controlled operators such as Telesp Celular have signed deals with Lucent and Motorola to develop 3G networks. Telesp, an 'A' band operator, does not intend to launch 3G services until 2008, but in the meantime intends to develop 2.5G at speeds up to 307 Kbps, launching a cdma2000 1xRTT network contracted from Lucent in Sao Paulo in December 2001.

The government of *Argentina* announced that it hoped to conduct an auction in October 2001. However, with the economy stagnating, the auction was delayed and this delay is likely to continue until there are real signs of economic recovery. Roughly $500 million is expected to be raised.

For its part, the government of *Uruguay* intended to auction 20-year tradable licences for both 2G and 3G spectrum in November 2001 using a simultaneous 'English auction' with complex rules concerning eligibility to bid. There are two 2G incumbents – state-owned fixed-wire incumbent Antel's subsidiary Ancel – operating using TDMA in the 1755–1770, 1850–1865, 1930–1945 and 2120–2135 MHz bands, and Movicom-BellSouth (BellSouth 46 per cent) – operating using CDMA in the 1890–1895 and 1970–1975 MHz bands. A 50 MHz block would be made available in 5 MHz paired lots (lots 1–5) in the 1800 MHz band (1710–1735 and 1805–1830 MHz). In addition, 50 MHz in 5 MHz unpaired lots in the 1900 MHz band (1865–1890 and 1945–1970 MHz) and 30 MHz in 5 MHz unpaired lots in the 2100 MHz band (2135–2160 MHz) (lots 6–10) would be available to be acquired as packages of either two or three lots covering both bands of which one lot must fall in the 1945–1970 MHz band. In these cases, if three-lot packages are acquired, residual 5 MHz lots may become available in a secondary auction, including the 2160–2165 MHz band on which Movicom-BellSouth has first option. The reason for this arrangement was to make it possible for bidders to acquire paired lots in the 1900 MHz band for PCS and paired lots in the 1900 and 2100 MHz bands for 3G. Operators would be able to acquire spectrum in all five bands subject to a cap of 15 MHz in each band and a total cap of 60 MHz. The choice of technology to be adopted would be left to licensees which would be able to use the spectrum for PCN, PCS and IMT-2000/ UMTS including TDD and FDD. Bidders were required to maintain a minimum level of activity in each round of the auction via either standing high bids, new high bids for any available lots or waivers. The licensees would be able to return the spectrum whenever they so wished and request that it be re-auctioned. The spectrum must normally be used within thirty months of the date of the licence. Sharing infrastructure is neither prohibited nor limited.[57] In November 2001, Lucent announced that it had been contracted to build a cdma2000 network for Movilnet, a wholly-owned subsidiary of CANTV, itself subject to a take-over bid. However, in the event, the government postponed the auction to 28 January 2002, fixing a reserve price of $6 million for each block of 5 MHz paired spectrum.

The situation in *Chile* remains ambiguous for now. Although it has the highest level of 2G penetration in South America – at over 30 per cent compared to 28 per cent in Venezuela – more than 75 per cent of subscriptions are prepaid so it is unclear whether customers are willing to pay large sums for sophisticated services. Furthermore, there is intense competition for what is, overall, a quite small market, suggesting that operators will be cautious about investing heavily in 3G. Entel PCS was the first to launch 2.5G with a GPRS service in October 2001, and its main competitor, Smartcom PCS, owned by energy group Enersis, is preparing to market cdma2000 1xRTT. Meanwhile, Telefónica Móviles and BellSouth are both hoping to secure an additional 10 MHz of spectrum during 2002 to permit the provision of 2.5/3G. No one expects 3G services to appear before 2005.

In Central America the government of *Honduras* announced a combined GSM and 3G auction to take place early in 2002. Of the five licences available, only one would be for 3G.

Meanwhile, in the Caribbean, *Jamaica* has announced its intention to auction two licences in early 2002.

Australasia

The auction of third generation spectrum in *New Zealand* had been scheduled for November 1999 but was subsequently cancelled because of fears concerning the inadequacy of the regulatory framework. Early in July 2000, the auction, combined with that for 2G batches of spectrum, was officially launched.[58] 3G licences providing management rights covering a 20-year period were on offer, each consisting of a maximum of 15 MHz paired in the 1920–1980 and 2110–2170 MHz bands, with one licence reserved for Maori interests (in the guise of the Maori spectrum holding company Hautaki Ltd) at a 5 per cent discount to the average price paid. The total spectrum of 60 MHz paired was split into 3 × 10 MHz paired and 6 × 5 MHz paired lots, and three additional lots of 5 MHz of TDD spectrum were also put on offer. Altogether, there were 20 registered bidders including Telecom New Zealand (TNZ), Vodafone Mobile NZ, Telstra Saturn, Walker Wireless, Broadcast Communications and Compass Communications. However, bidding was initially slow – partly because of the limited size of the overall market – especially for 3G spectrum where incumbents TNZ, Telstra Saturn and Vodafone Mobile NZ predictably did just enough to keep control of one licence each via minimum incremental bids in each round.

The bidding was initially supposed to end when two consecutive rounds occurred with neither valid bids nor withdrawals of existing bids. A clear round first occurred at Round 191 after three months of bidding, subsequent upon the withdrawal of Walker Wireless and Independent Newspapers, at which point Telstra Saturn had bid $8 million, TNZ $7.8 million and Vodafone Mobile NZ $7.2 million – very modest sums in relation to those offered elsewhere.

In round 207, Clear Communications (owned at the time by BT) made a belated entry into the auction, bidding for both 3G and 2G spectrum, and by round 240 was dominating the 3G bidding. However, bidding dragged on so slowly that in late November the government altered the rules such that the auction would end after a single clear round, with the number of rounds per day reduced from six to four. On 8 December 2000, the auction was halted after 380 rounds because bidders claimed that the value of the spectrum would be affected by regulatory changes about to be introduced by the government, and a resumption date of 8 January 2001 was set. At that point the 3G management rights (including those covering the FDD and TDD lots) were separated from the 2G licence/1098 Plan licence lots with both auctions set to end independently after one clear round. It was also ordained that if a bidder withdrew its bid for management rights in five separate rounds it would be excluded from making new bids although existing ones would remain on the table, and that no new bidders would be permitted to register.

The management rights auction came to a close in mid-January when there were no bids made during round 415. TNZ bought 10 MHz paired plus 5 MHz paired for $7.5 million; Vodafone Mobile NZ bought 10 MHz paired for $4.6 million and 5 MHz of TDD for $0.75 million; Clear Communications bought 10 MHz paired for $4.5 million and 5 MHz of TDD for $0.80 million; and Telstra Saturn bought 2 × 5 MHz paired for $3.7 billion and 5 MHz of TDD for $0.75 million. Technically, only TNZ has enough spectrum for nation-wide coverage, officially estimated to require 15 MHz paired, but the other three licensees may well believe that they can achieve this with somewhat less. Why Telstra Saturn only bought 5 MHz lots, and why they were so cheap, remains a mystery.

In September 2001, start-up mobile operator Econet Wireless NZ Ltd, a subsidiary of Zimbabwe-based Econet Wireless International, announced a deal to utilise the 2G and 3G spectrum set aside for Maori interests. Hautaki Ltd agreed to exchange access to its 3G spectrum over a 10-year period for the right to take a 30 per cent stake in Econet Wireless NZ – the price agreed in January 2002 was $1.7 million with Econet paying Hautaki $25 000 a year for Maori scholarships – although reservations were expressed by analysts about Econet's ability to finance a network roll-out. By way of assistance, the government agreed to introduce a rule requiring mobile operators to share transmitter sites and to grant roaming rights. However, as a result of protests by the incumbents, the government subsequently stated that the new rules for roaming would require new entrants to achieve 10 per cent national coverage before roaming rights would be mandatory. Econet has arranged to share Vodafone's 2G network.

In November 2001, Telstra Saturn agreed to buy Clear Communications from BT for $165 million. As a result, the newly formed Telstra Clear found itself with 20 MHz paired of spectrum, exceeding the cap by 5 MHz paired which will now either have to be returned to the government or sold on to another operator within three months. The government also retains a block of 5 MHz paired left over from the first auction which it will sell during 2002.

Meanwhile, the *Australian* government announced that it would be auctioning 15-year licences, operational from October 2002, in mid-March 2001. A simultaneous multi-round auction would be utilised. The available spectrum in the 2 GHz band, which covers 95 per cent of the population, was divided up such that 45 MHz paired plus 20 MHz unpaired was on offer in Canberra; 60 MHz paired plus 20 MHz unpaired was on offer in other state and territory capitals (in both cases with a metropolitan bidder's cap of 15 MHz paired and 5 MHz unpaired); and

20 MHz paired was on offer elsewhere (with a regional bidder's cap of 10 MHz paired). The 58 lots were arranged such that two national licences were available with a further two made possible by aggregating up the requisite lots. A notional value of $1.6 billion was placed on the licences, but the New Zealand experience, not to mention the fact that all the major operators had announced that they would not pay 'silly prices', with One.Tel refusing to bid altogether, indicated that this was unlikely to be anywhere near achieved. Registered bidders comprised the 2G incumbents bidding as Telstra 3G Spectrum, C&W Optus Mobile and Vodafone Pacific, fixed-wire operators bidding as Hutchison Telecommunications (Australia) (Hutchison Whampoa 58 per cent) and AAPT Spectrum (Telecom New Zealand) and newcomers CKW Wireless (a subsidiary of US-based ArrayComm) and 3G Investments (Australia) (Qualcomm). The 2G incumbents plus 3G Investments made the maximum down-payment, indicating a wish to bid for a national licence, whereas Hutchison and AAPT bid slightly less and CKW a very small amount.[59] Somewhat surprisingly, AAPT withdrew in mid-March.

On the first day of the auction, 32 of the 58 lots attracted bids. On day four, the aggregate reserve price was almost reached with Telstra, Vodafone and C&W Optus pursuing the two national licences. The auction ended after six days and nineteen rounds with 48 of the lots under offer. The national licences went to C&W Optus, which also bought other spectrum and paid $123 million, and Vodafone Pacific which acquired slightly more spectrum than Optus and paid $126 million. However, it was Telstra that paid the most – roughly $150 million – to buy more than enough spectrum for full national coverage. In addition, Hutchison paid $97 million for spectrum covering the five largest cities, 3G Investments acquired 10 MHz paired in the capital cities for $79 million and CKW bought 5 MHz unpaired in the same cities for under $5 million which it intends to use exclusively for data services. Altogether, just over $580 million was raised, little more than the reserve and well below initial forecasts.

In May 2001, Telecom New Zealand announced strategic alliances to develop 3G services with both Hutchison Whampoa – via its subsidiary Hutchison Telecommunications (Australia) – and Microsoft. In the former case, the joint venture would be called Hutchison 3G Australia (H3GA) in Australia – with TNZ acquiring a 19.9 per cent stake – and Telecom 3G in New Zealand – with Hutchison acquiring an option to take a similar stake. In both cases, the Hutchison brand would be used. Further joint ventures could follow. Meanwhile, Microsoft would invest $127 million in TNZ and combine its MSN New Zealand consumer portal with TNZ's Xtra portal to create XtraMSN. In January 2002, H3GA announced that it would be acquiring in April most of the 2G assets built, and later reclaimed, by Lucent for the now bankrupt One.Tel. This was claimed to provide H3GA with 60 per cent of the base stations it would need to roll out 3G, permitting services to be launched in Sydney, Melbourne and Brisbane possibly as early as the end of 2002.

The Middle East and Africa

Israel, where mobile penetration is unusually high at over 70 per cent, intended to launch an auction with a closing date for submission of applications of July 2001. Four licences were on offer, each initially at a minimum price of $100 million. Each licence would operate on a separate frequency, although it was unclear when the relevant frequencies would be available on an unencumbered basis. The four incumbents – PelePhone Communications (Bezeq Telecom 50 per cent, US-based Shamrock Investments 50 per cent), CellCom Israel (backed by BellSouth), Partner Communications (Hutchison Telecom/Matav Cable Systems – the only GSM network) and newly licensed MIRS Communications were expected to be the only bidders. PelePhone proposed to use Motorola technology to launch 3G services at 144 Kbps in early 2002, but was unhappy at what it saw as the high cost of a licence and was undecided whether to utilise existing spectrum. The view was widely held that, in practice, carriers would concentrate upon 2.5G until 2004. It was alleged in May 2001 that Ericsson had offered to build and finance a network that could be shared by several operators, and network sharing in some form was seen as inevitable. In August, minimum payment terms were set at $52 million overall per licence, of which half was to be paid in 2002 and the rest by 2006. PelePhone Communications responded with the threat of an appeal on the grounds that the government had previously agreed to accept royalties based upon revenues, but in the event, when the licences were awarded in December 2001, PelePhone proved willing to pay the minimum fee as did CellCom Israel and Partner Communications. In addition, CellCom and Partner paid $42 million for extra spectrum to run 2.5G services. In January 2002, Partner Communications announced that both its 2G and 3G licences had been extended by nine years to 2017.

In *Turkey*, the incumbents Turkcell and Telsim had the 2G market to themselves until new PCN licences were awarded to Aria (part-owned by Telecom Italia) and fixed-wire incumbent Turk Telecom (trading as Aycell) to commence services in April 2001. However, despite the relative lack of development of GSM, Telsim signed an agreement with Motorola for the supply and installation of a 3G network in October 2000. The deal involved $2 billion of vendor finance, but, in April 2001 Telsim failed to make a repayment of $728 million and Motorola threatened a formal default in July if no progress was made. Motorola is entitled to either receive or sell 66 per cent of Telsim stock in the event of a default. The licences are now likely to be offered at the end of 2002, with the number of licences determined by the number of bidders, but no rules have yet been published

The situation in the *United Arab Emirates (UAE)* is rather odd because, although there have been no official announcement on the matter, GSM incumbent Etisalat is apparently building a 3G network which the government expects to come into service by the end of 2002 – making it possibly the next after DoCoMo to launch commercially.

In *South Africa*, the proposal contained in the recent draft telecoms policy statement is to award five licences to fixed/mobile incumbent Telkom, the about-to-be-licensed second national operator and the three other mobile incumbents – Vodafone's Vodacom, M-Cell's MTN and new entrant Saudi-backed Cell-C – all of which would pay an 'appropriate' fee.

Operators' interests in 3G licences and measures of licence costs

Table 4.2 below indicates the position in March 2002 in respect of operators' interests in licences awarded up to that point in time. Their proportionate stakes have been set out in the preceding material and are constantly evolving.

The issue of how much has been paid in each country on a comparative basis is somewhat difficult to resolve, in part because countries vary in the number of licences issued and the licences do not provide a uniform amount of bandwidth. An attempt has been made by the UMTS Forum (see www.UMTS-Forum.org) to provide a consistent measure expressed in terms of Euros per pop and per 2×5 MHz of spectrum. This shows, first, that the cost per 2×5 MHz has varied so far between €3 million in New Zealand to €4270 million in Germany with only the UK (€3543 million) and Italy (€1224 million) also running into four figures compared to thirteen cases where the figure remains in two figures and five (France, Spain, South Korea, Netherlands and Poland in descending order of magnitude) where it runs to three. Once the data are adjusted for population size the UK comes out slightly higher than Germany

Table 4.2 Operators'[1] interests in 3G licences by country

	\$ per pop[2]	Belgacom	mmO_2	Deutsche Telekom	DoCoMo	Hutchison	KPN	Orange	Sonera	TDC	Telecom Italia	Telefónica	Telenor	Tele2	Telia	Vodafone
Australia	30					L										L
Austria	90			L		L		L		L	L	L	L			L
Belgium	45[3]	L			L		L	L								L
Czech Rep.	27			L			L			L	L					
Denmark	85					L		L		L				M	L	
Finland	neg.								L					L	L	
France	140[3]		L					L								L
Germany	560		L	L		W	L	L	L			L				L
Greece	40[3]							L			L		L			L
Hong Kong	35–55		L		L	L										
Israel	30					L										
Italy	180					L		L	L		L	L				L
Japan	0				L											L
Netherlands	159	L	L	L	L		L	L		L				M		L
New Zealand	neg.		L[4]			O										L
Norway	neg.								A				L	L	L	
Poland	45			L				L		L						L
Portugal	35							L					L			L
Singapore	40		L		L											
South Korea	60		L													
Spain	10								L			L				L
Sweden	neg.					L		L						L	M	L
Switzerland	15							L		L		L				L
UK	585		L	L	L	L	L	L								L
in USA?				L	L											L

Notes

1 The licence is often held by the majority-owned mobile subsidiary as in the case of Telecom Italia Mobile, but in certain cases that subsidiary has been hived off as in the case of mmO_2, formerly BT Cellnet. Hutchison's licences are held by a variety of separate subsidiaries.

2 The total sum raised divided by population as an approximation. It can be argued that these figures are deceptive because they do not take account of the number of licences issued – that is, the number of potential subscribers available on average per licensee. This has the effect of raising the German figure (\$560 × 6) above that for the UK (\$580 × 5).

3 The total sum raised in these cases applies to licences so far awarded and is subject to alteration.

4 BT has agreed to sell its licensee, Clear Communications, to another licensee, Telstra Saturn.

neg. = negligible; n/a = not available; A = abandoned; L = licensee; M = MVNO; O = option to take a stake in licensee; W = withdrawn from licensee.

(€59.5 per pop/2 × 5 MHz compared to €51.6), with a huge gap to Italy in third place followed by Slovenia, the Netherlands, Spain and France. New Zealand remains much the cheapest at €0.8 followed by Switzerland.

All such measures have to be taken with a pinch of statistical salt but the UMTS Forum believes that two conclusions can reasonably be drawn, namely that there is, first, a clear tendency for the price of spectrum to decline over time reflecting a deteriorating business climate and, second, a tendency for the price to be positively related to the size of the market, especially if GDP is used rather than population as the measure of size. The Forum observes finally that the effect of auctions in big countries has been to shut out local players completely, and in small countries to shut them out most of the time. In contrast, beauty contests have allowed local players to obtain one third of the licences in all sizes of country which they interpret as providing a good reason for beauty contests to be preferred to auctions. As the extensive discussion of this issue in the next chapter reveals, there is not much evidence to support this contention, and, in any event, the (unexplained) division into local, regional and global players appears to be rather arbitrary if one inspects the make-up of licensees in the material above.

5
Analysis of Issues Raised by 3G

Strategic partnerships

Attempts to consolidate incumbents in Europe have so far proved to be extremely unsuccessful: Deutsche Telekom failed to take over Telecom Italia; Telia and Telenor failed to merge; Telefónica failed to win KPN; and TDC failed to merge with Telenor – and these were only the tip of the iceberg because, for example, attempts other than via take-overs to link up the likes of BT with Telefónica and Deutsche Telekom with France Télécom also fell apart.

However, the movement towards consolidation appears not to have died on its feet but to have transferred to the mobile sector, and when all of the major incumbents have completed an initial public offer of their mobile subsidiaries – assuming market conditions permit – we must surely expect some further linkages in addition to those being promoted by the pursuit of 3G licences.

Meanwhile, as noted above, relatively few licence bids have been made so far by individual companies, partly as a consequence of the UK experience which indicated that the outlay would be much higher than originally forecast in each country. Furthermore, there were obvious advantages in obtaining licences in as many countries as possible, which greatly increased the aggregate outlay needed to gain licences throughout Europe.

This led commentators to conclude that a small number of bidders, mainly joint ventures, would rapidly come to dominate the bidding. This was borne out in particular by the fallout from TIW's successful bid in the UK. Despite being cash-rich, Hutchison Whampoa was understandably reluctant to foot the majority of the bill for this and other bids, and within the remarkably short period of four weeks created the so-called 'golden triangle' with KPN and NTT DoCoMo. DoCoMo initially took a 15 per cent stake in KPN Mobile. The rest of the relationship was country-dependent as follows:

UK: DoCoMo bought a 20 per cent stake in a new licence-holding venture, Hutchison 3G UK Holdings – which was approved by the European Commission in September 2000 – for $1.8 billion and KPN Mobile bought a 15 per cent stake for $1.35 billion. TIW was authorised to buy a further 6.5 per cent until November 2000, but declined, leaving Hutchison with 65 per cent.

Germany: Hutchison and KPN Mobile formed a new company called E-Plus Hutchison as the vehicle for a bid. The intention was that if the bid was successful, they would jointly build the network and then use it as competing carriers, leaving each free to float its own operation as and when it so desired. In practice, as soon as the spectrum was won at a cost of $7.62 billion, Hutchison withdrew, claiming that the joint bidders really needed three blocks between them, and transferred its share to E-Plus (now owned by KPN Mobile 77.5 per cent; KPN Telecom 22.5 per cent).

France: The three carriers intended to bid together, preferably with a French equity partner, and if successful to mount a joint operation.

Belgium: Hutchison and KPN Mobile intended to bid jointly.

From KPN's viewpoint the arrangement meant that its lack of funds would not hold it back as it had done in the UK auction. Hutchison, meanwhile, was being paid $3.15 billion for 35 per cent of a UK licence costing $6.7 billion – a premium of roughly 37 per cent. Why DoCoMo was taking a minority stake was somewhat less clear given its resources, but it was thought to want to restrict its main efforts to developing its i-mode service.

Hutchison's withdrawal from the agreed arrangements in Germany was allegedly irrelevant for arrangements elsewhere. However, KPN's response was to express severe reservations about whether it wished to proceed in France where it had failed to seal an alliance with either Vivendi or Bouygues Télécom. It was also considering whether to proceed by itself in Belgium. For its part, Hutchison indicated that it did not wish to bid in France, ostensibly because it could not find a suitable partner.

In January 2001, Telecom Italia Mobile (TIM), which had previously had no partnership arrangements with other European operators, joined up with KPN Mobile and DoCoMo to develop i-mode based services for the European market, commencing with Germany, Italy, the Netherlands and Belgium. The proposal was to launch services via a handset offering i-mode overlaid on an existing WAP service using new GPRS-enabled networks. The companies have yet to show any inclination to act as a stable consortium, but it seems probable that in some combination they will remain a force in the 3G arena.

In addition to the above, it was widely agreed that a maximum of a further four core bidders were likely to emerge with pan-European reach. Either independently or via joint ventures, a central role would be played by the Vodafone Group, Deutsche Telekom and Orange (France Télécom), with BT a less plausible member of this select group given its many minority 2G stakes. It was notable, for example, that each of the four incumbents in the Netherlands other than KPN was backed by one of these four.

However, none of the above arrangements involved Telefónica which became steadily more active in setting up consortia in conjunction with Sonera and Suez Lyonnaise des Eaux (now Suez). With the licences already awarded in Italy, Suez paid $90 million for a three per cent stake in IPSE 2000 which it bought from Telefónica, with a clause enforcing the payment of damages if either party reneged on the deal. However, the joint venture with Suez failed to submit a bid in France, and the joint ventures with Sonera in Germany (Group 3G) and Italy (IPSE) soon came into question when Sonera, desperate to resolve its spiralling indebtedness, refused to invest any more than was absolutely necessary to meet contractual obligations. Indeed, Sonera's joint ventures have in general been fairly disastrous, with the bankruptcy of Broadband Mobile in Norway and the failure of Xfera to obtain a GSM licence in Spain to be added to its other misadventures.

Meanwhile, as noted above, Hutchison Whampoa has begun to create additional joint ventures both within Europe – notably in Sweden – and elsewhere in pursuit of its strategy of partnering leading operators wherever it is active – in the latter case with Telecom New Zealand covering both Australia and New Zealand. It is intended that TNZ will acquire a 19.9 per cent stake in the new company, capitalised at one billion Australian dollars. Now that licences have been awarded elsewhere in the Asia-Pacific region – for example, in Hong Kong – no doubt other joint ventures will follow.

Auctions versus 'beauty contests'

Historically, radio spectrum in general has been viewed as a scarce natural resource which, for reasons of efficiency and equity, has needed to be rationed by governments and/or international agencies.[60] However, demand for spectrum has tended to concentrate upon particular wavelengths which have become increasingly valuable, while others have tended to be treated in a somewhat profligate manner because at the time of their allocation there did not appear to be any use of much value to which they could be put. Although technological advances have made it possible to reduce interference between wavelengths, permitting more efficient use – and hence expanding the supply – of suitable spectrum, the recent rapid development of mobile telephony has nevertheless transformed the balance between the supply and demand of suitable spectrum and forced governments to reappraise how this should be allocated and, in particular, whether to involve markets in the process. The main objectives to be satisfied in allocating licences to use spectrum are generally agreed to be promoting a competitive industry, making licences available to whomever will make best use of them and, through the determination of the appropriate value to be put upon a public asset in the form of spectrum, raising some revenue for the government in a non-distorting fashion. But what method of allocation best meets these objectives?

Although they had some prior history in New Zealand, the use of auctions to allocate spectrum mainly originated in the USA where bureaucratic systems of allocation – generally known as 'beauty contests' but as 'comparative bidding' by the European Commission – went out of fashion in the late 1980s. Prior to this time wireless licences for what were known as Personal Communication Services (PCS) were assigned by the FCC subsequent upon administrative hearings during which prospective licensees sought to persuade the FCC of the value of the uses to which they wished to put the spectrum. However, the length of the hearings and the associated backlog of unallocated licences, the lack of transparency of the proceedings and the failure to raise any revenue led Congress to switch in 1982 to a process of allocation by lottery.

Although this was a speedier process, it resulted in applications by parties whose intent was not to use the licences but to sell them on at a profit.[61] In mid-1993, Congress grew tired of handing over money to speculators and asked the FCC to determine an optimal method for auctioning spectrum. With little prior experience to fall back on,[62] the FCC issued a 'Notice of Proposed Rule Making' in September that set out to achieve an efficient solution while simultaneously raising revenue for the government, preventing the acquisition of monopoly power and ensuring that smaller organisations outside the larger conurbations were not disadvantaged.

The process of designing the rule book for the auction that eventually took place in July 1994 involved theorising, experimenting and simulations, and was notable for its recourse to economic game theory. Because this and the ensuing auctions through to the end of 1997 raised $23 billion for the government, they certainly succeeded in meeting their primary objective. In other respects, however, they were less successful, finding it hard, for example, to address the problem that if licences are offered in adjoining areas, the value of each licence is partly determined by whether the bidders have already acquired, or hope to acquire, adjoining licences. The 1993 Rule Making Proposal tried to deal with this with a proposal that involved first offering the licences as a complete set in a sealed-bid auction, and then individually. Whichever system produced the higher revenue would ultimately prevail.[63]

Not everyone was convinced of the desirability of the FCC approach, and an alternative was tabled in the form of a 'simultaneous ascending-bid auction' (SABA). The key properties of such an auction are that bids can be made from among a choice of licences, or licence packages, on offer at the same time, and that new bids will continue to be accepted so long as they exceed those already tabled. Crucially, this allows bidders to switch from targeting one particular licence, or set of licences, to another as bidding proceeds, taking account of all bids previously tabled by competitors. In addition, because the existing level of bids is known, there is no need for a determined bidder to table a hugely inflated initial bid for fear of making an unsuccessful offer – the so-called 'winner's curse' associated with sealed-bid auctions.

There was an understandable reluctance to introduce an unproven procedure which, compared to previous methods, was relatively complex,[64] and, at the end of the day, it does have to be recognised that it is not possible to determine whether auction outcomes are optimal either purely by recourse to theory or, indeed, by studying outcomes of actual auctions where the thought processes of bidders are not available for analysis. The sensible approach is accordingly to engage in experimentation to reveal where theory and experience fail to marry up, and to improve the rules in stages.[65]

The process of experimentation broadly supported the conclusion that the SABA would be more efficient than the combination auction proposed by the FCC. However, a practical rulebook has to address a wide range of issues including eligibility to bid, the right to withdraw, the right to pass on a round of bidding, the size of bid increments, the number of rounds per day and so forth. A standard variant of SABA evolved such that all prospective bidders would initially become eligible to bid by paying an entry fee, where the number of entry fees would reflect the number of independent licences being sought. Valid bids

GAME THEORY.

d then be made known to all par-
ne 'standing high bid' and invited
at a minimum percentage incre-
e bid would require an eligible
to the standing high bid plus
ned to be 'active' if they either
made a subsequent eligible bid.
n active in respect of a specified
were eligible to bid, they would
in subsequent rounds – thereby
ds early on, and hence increas-
ait and see' behaviour. Bidders
row constraints to avoid serial
ndraw bids. In the latter case,
failure to acquire a particular
ng bidder would be obliged to
withdrawn bid and the price

etween openness and privacy.
ch more open than the sealed-
nportant element of privacy –
g place – in that the winning
was ultimately prepared to pay.
ween increments and speed.
ction with lots of eligible bid-
hundreds of rounds. On the other hand, the imposition of large increments designed to speed up the process may cause bidders to drop out too quickly, with a potential loss of revenue to the auctioneer. This is fostered by the fact that a lengthy auction leaves time for bidders to return to their shareholders and/or the financial markets to seek permission to commit additional monies to the auction. Given the existence of this and other trade-offs, it is understandable that there is no definitively optimum way to design an auction, and hence, in practice, no two are ever exactly alike.

For its part, the European Commission initially requested member states to award UMTS licences without charge, a policy adopted when the first licences were awarded in Finland. This reflected the fact that construction of a network was expected to cost upwards of $3 billion with annual running costs of $750 000 in a country the size of the UK. However, the general scarcity of spectrum meant that it was an attractive prospect for governments either to conduct 'beauty contests' – where the licences are awarded to whichever applicants make out the best case for use of the available spectrum in return for a fee or royalties – or to auction licences, since both methods would ensure at least some income for the state.

It must be borne in mind that the scarcity of the spectrum to be made available was bound to have an effect upon the market value of the licences, and so far there has been what some commentators feel is a curious reluctance to utilise the full range of spectrum suitable for 3G.[66] Whether, and how, additional spectrum would be made available at some later date was ultimately a country-specific issue, and it was unusual in practice for anything to be said in the licensing conditions about this matter. Furthermore, it was unusual for there to be any clear official statement about the possibility of trading spectrum rights after licences were awarded. This meant that potential bidders were taking a major risk if they chose to stay away from an initial auction in the hope either of making a bid on a later occasion or buying rights from licensees.

The obvious danger with auctions was that the money paid to the government would no longer be available for investment in the networks – a variant of the 'winner's curse' – but governments were only too well aware that a properly conducted auction could potentially work wonders for government finances. The crucial issue was perceived to be the specifics of the auction process which needed to be tailored to the particular circumstances at hand. This had become clear, for example, during the 1995 wireless auction in the USA when an incumbent, which had the most to lose because it was the main fixed-wire carrier, and which also possessed the biggest database of subscribers, was able to win with a very low initial bid because potential competitors assumed from the off that it would do whatever was necessary to win and hence chose not to participate. Another feature of the same auction was that companies were allowed to bid for multiple licences. This meant that companies had an incentive to win adjoining licences, and hence to signal their intentions to other companies in the hope that they would collude in carving up the market.

For once, economists – and especially game theorists – seem to have emerged with their reputations enhanced. The early UK auction was very carefully designed to achieve all of the cited objectives. Not only were there more licences on offer than incumbents – five compared to four – but the largest licence was reserved for a new entrant thereby providing an incentive for a considerable number to bid.[67] Secondly, each participant could only bid for a single licence through a series of rounds during which all bids were placed simultaneously. Hence, at least one newcomer had to be left in the bidding at all times. Furthermore, since one licence was set aside exclusively for a new licensee, if bidding for that licence grew too intense then new entrants had an incentive to switch to other licences being contested by incumbents – which would therefore be unable to win these by selecting one each, putting in a low bid, and refusing to compete with one another. If nothing else, the amount raised in the UK auction demonstrated that there was no collusion of any consequence.

In common with other auctions in the USA, the amounts bid were made public rather than hidden in sealed bids. This meant that if one licence was going relatively cheaply it would attract additional bids, and all licences with the same amount of spectrum on offer could expect to attract almost identical bids in any one round. Each potential bidder would feel free to switch licences during the early stages, and hence to bid tactically and disguise their intentions, but as the bidders left in the auction fell close to the number of licences on offer they would be forced to reveal their hands, since the instant that there were only five bidders left, each would have to accept whichever licence they held the highest bid for at the time.

The UK and most other European auctions used pay-your-bid pricing whereby each licence is sold at the specific price offered by the successful bidder. The alternative is to use a uniform pricing method whereby every licence is sold at the same price, possibly that of the lowest successful bid or the highest unsuccessful bid. However, whereas these methods would potentially produce significantly different outcomes under a sealed-bid auction, this does not tend to be the case for ascending-bid auctions since bidders have little incentive to raise their offers by more than the minimum increment required by the rules.

As noted, it is always possible to argue that bidding rules can never emulate precisely the outcomes that would come about in free markets.[68] Nevertheless, it should be possible to learn even from the experience of slightly flawed auctions and it does seem somewhat surprising that subsequent auctions did not introduce some of the features that had made the UK auction such a success – at least from the government's point of view. For example, the auctions in Germany and the Netherlands failed to provide for a guaranteed new entrant. Analysts in Germany quickly noted that if all, or even most, of the incumbents bid for the maximum spectrum available then there would be none left for new entrants – and hence incumbents had an incentive to shut new entrants out and new entrants had an incentive to link up with one of the incumbents. Equally, in the Netherlands, with the number of licences equal to the number of incumbents, the sensible strategy seemed to be to link up with one of them rather than bid against them. Hence, the notion of real competition seemed unrealisable. This was accentuated in the case of the Netherlands where bidders chose not to bid in every round during the early stages – a particularly effective tactic since the auction rules stipulated that reserve prices would be reduced every time there was no bid.

That so much money was nevertheless raised in the German auction was argued to be more to do with irrationality on the part of participants than with the design of the auction. As noted above, the incumbents began by driving up prices while bidding for the maximum permitted amount of spectrum, only to allow the auction to end abruptly with all bidders getting an equal share – apparently the worst of all possible outcomes from the incumbents' point of view. This view is contested on the grounds that the German auction was an improvement upon that in the UK insofar that it permitted bidders to determine for themselves how many spectrum blocks to pursue. Had the UK followed this principle, it might have ended up with six licensees, more competition and more revenue. This is because, where the number of licensees is initially unknown, all potential bidders know that there is some chance of obtaining a licence, whereas potential bidders rapidly learn from a UK-style auction that incumbents will win any licences available to them and hence non-incumbents will withdraw elsewhere (as in Switzerland).[69]

Where some auctions – for example in the Netherlands – also clearly went wrong was to rely excessively upon the multiple-round ascending 'English' auction, which worked perfectly well in the UK because there were so many bidders and more licences on offer than incumbents. These auctions should have been concluded with a sealed 'best and final' so-called 'Dutch' offer while there was still one more active bidder than the number of licences. This is because if a new entrant makes a higher 'best and final' bid than an incumbent, the latter cannot return to bid even more and must accordingly itself bid a realistic price as a precaution. Furthermore, the forming of consortia is discouraged because even new entrants believe that they have a 'sporting chance' of acquiring a licence while bidding independently.

Under the circumstances, it was somewhat surprising that the Italian government went ahead with its auction with only one participant more than the number of licences available, especially given the fact that one bidder was clearly riven by internal disputes. The government was extremely upset when this bidder withdrew after only eleven rounds of bidding, bringing the auction to a sudden halt, but given that it had itself drawn up the rules it had no recourse other than to claim that collusion could have taken place to keep down the price of licences, even though hardly anyone really believed this to be true.

The evidence tends to support the view that the German auction would have performed better had the spectrum been allocated to specified licences in advance, as in the UK and elsewhere. Bidders appear to have strong preferences for specific spectrum, and hence there is likely to be a major dispute with the regulator if, as is likely, they do not initially each choose a different band when the auction is concluded.

One of the obvious difficulties with conducting auctions in sequence is that potential bidders do indeed learn from experience. In particular, they tend to form consortia to ensure that they have at least a partial stake in a licence – and the opportunity to renegotiate those stakes at a later date. Once it was realised after the initial auctions that consortia were being formed in most other countries and that this was going to reduce significantly the amounts being bid, governments had several options open to them other than to do nothing. They could have redesigned the auctions and/or introduced much higher fees and/or introduced rules to prevent collusive behaviour. All remedies had their merits, but above all it was desirable to prevent the bidders extracting monopoly profits. In this respect there was an interesting contrast between the UK case where assets were shuffled around – for example, the sale of Orange – *after* the auction was concluded, and the case in the Netherlands where new entrants and incumbents resolved their participation in consortia bids *before* the auction took place.

The behaviour of the Spanish government was another matter altogether. Having opted for a 'beauty contest', and awarded licences for a modest up-front fee plus a small annual charge in order to exert pressure upon licensees to make large and speedy investments, it came to regret both moving early and its method when it saw how much money had been raised in the UK and Germany. Its decision to increase retrospectively the annual charge may have been technically legal, but it was certainly morally dubious. The Italian government did at least have the grace to accept that it was responsible for the rules and hence that the original licences rules should stand. On the other hand, the investigations launched by the magistrates in Rome and the antitrust authorities into 'irregular' behaviour, not to mention the action of the government in demanding that Blu forfeit its deposit, indicated that 'sour grapes' were widely in evidence in official circles.

So was there indeed a 'winner's curse' associated with the auctions? It was widely believed that this was the case in the UK and Germany because the bids eventually rose to a level that suggested the winning bidders would lose money if they used the licences to build new networks. Perhaps so, came the reply, but it had to be borne in mind that participants had plenty of time to work out in advance exactly what they were willing to pay, and knew throughout the auction exactly how much each unit of spectrum would cost and hence exactly what it was worth to other well-informed bidders. If, indeed, they paid too much – and only time will tell – they therefore had nobody to curse but themselves.[70]

Part of the difficulty lay in determining whether high licence fees would necessarily result in higher prices for services compared to where licences were cheap. It was argued, on the one hand, that this was not logical since, in principle, licence fees, which are sunk costs, should not affect pricing decisions which are based upon a comparison between marginal revenue and marginal cost. Indeed, high fees would provide a spur to a rapid roll-out of networks in order to recoup the investment as quickly as possible. It was pointed out, on the other hand, that in practice all licensees would have an interest in passing on any costs they could, and insofar that they were successful in passing on licence fees in higher prices, there would be a consequent downward pressure on demand where auctions resulted in high fees. This would mean, in turn, that the objectives discussed above are inherently contradictory, and that an auction cannot bring about an optimum industrial structure from the point of view of public welfare.[71] Furthermore, outcomes would be very different as between countries adopting auctions and those relying on 'beauty contests'.

If nothing else, the above discussion demonstrates that auctions tend to produce suboptimal outcomes unless they are designed with great care, and for that reason alone many analysts prefer to rely on 'beauty contests'. However, letting civil servants make choices via 'beauty contests' does not strike everyone as a good way to allocate licences since it implies that they have judgmental powers that in some mysterious, non-transparent manner, bring about outcomes preferable to those brought about by free markets. Not that auctions are themselves necessarily entirely non-political since, as shown above, there may be restrictive conditions placed upon bidders to, for example, create consortia encompassing domestic companies.

A further factor is that if spectrum is free or extremely cheap, there is an incentive for companies to acquire and hoard it until such time – if ever – that they can put it to good use, in the process denying it to other companies which develop potentially valuable uses for the spectrum at a later date.

Interestingly, and, perhaps, confusingly, the fact that some auctions produced very high fees while others produced relatively little, and that some 'beauty contests' involved minimal fees while others involved fees somewhat higher than the sums raised in many auctions, means that it will never be possible to distinguish between the two approaches with any real degree of certainty and, indeed, this uncertainty was compounded in the case of France by the alteration in the structure of the licence fees after the initial tender failed to secure the required number of applicants.

Perhaps, therefore, the crucial lesson is simply that, whichever method is chosen, it should be chosen with great care with a clear view as to what can reasonably be achieved at the time the licences are issued. That much the French government has learnt to its cost. In the first place, the government, by no means untypically, tried to ensure that all the incumbents would get a licence, and that they would pay a modest amount compared to the fees achieved specifically in the UK and Germany. This would hopefully ensure that they would not face a cash crisis and hence become vulnerable to a take-over by a foreign operator. Secondly, the government wanted to lay down strict rules on quality even though these would necessarily depress the perceived values of the licences. Finally, the government promised in advance to use the cash raised for its pension fund, and hence did not want to scale back the fees once set. By refusing to be flexible in the face of the rapidly deteriorating value placed upon 3G licences, the government ended up selling only two, thereby creating a short-term duopoly, ensuring that non-incumbents would refuse to participate even if the rules were changed, raising much less money than expected and making itself look incompetent.

Sources of finance

A further factor that distinguishes the two approaches concerns the need to raise financing for 3G networks. In principle, the amount of money that a licensee needs to earn overall must at least cover financing costs. Since auctions typically – albeit not necessarily – involve higher costs than 'beauty contests', the resultant charges for 3G services must in the end be higher on average if losses are to be avoided, although it is argued in some quarters that competitive pressures will ensure that the difference will not be significant at any point in time and hence that the profit margin will simply end up generally lower for licensees participating in auctions.

Licensees face two financing issues. The first is the need to finance up-front licence fees, and the second is the need to finance network roll-out. The first issue is more problematic than the second since it is more immediate and roll-out can generally be part-financed using cash flows from existing 2G services.

The licence costs are set out in Tables 3.1 and 4.1. As can be seen, these bear no common relationship to the size of the country in question. However, such a relationship can broadly be expected to apply in respect of roll-out costs although some differences arise because the population is much more dispersed in some countries compared to others. How significant this is depends upon the percentage of the population that licensees must reach either because of a licence condition or a voluntary offer made in the course of a 'beauty contest'. In the case of Sweden, for example, the up-front licence fee was nominal but the offered coverage was effectively 100 per cent and the most isolated customers will clearly be very expensive to reach. The largest individual roll-out cost items are base stations. In urban areas there are economies of scale to be derived from constructing these as large as possible, but in recent times there has been an environmental backlash against tall masts and prime rooftop sites are becoming increasingly scarce and expensive. In addition, licensees must account for handset subsidies, customer acquisition, marketing expenses, content acquisition and product development.[72]

Not all licensees are cash-strapped – Vodafone, for example, is not expected to be overly troubled by the demands made by its 3G networks – but the collapse of confidence in the telecommunications sector during the latter part of 2000 – partly itself a reflection of spiralling 3G commitments – left most operators with significant financing problems. These could, in principle, be approached in four ways, namely the issue of new equity, the issue of commercial paper and bonds, bank loans and equipment vendor finance. However, whereas the first option was almost always to be preferred, it was not necessarily feasible because the stock markets were depressed. Thus, although Telefónica Móviles and KPN did succeed in floating new shares in the latter part of 2000, these had to be sold at rock-bottom prices which even then were difficult to sustain. France Télécom faced a similar situation when floating a stake in Orange – into which it had rolled up its various mobile interests – in February 2001.

Although Deutsche Telekom also planned to issue shares in its mobile subsidiary during 2001, it failed to do so and, even if successful in 2002, is unlikely thereby to be able to raise the sums required for its 3G licences and networks. This has meant that a great deal of recourse has needed to be made to the bond markets. For example, BT raised $10 billion in dollar-denominated bonds. However, not only was the interest rate on offer much higher than customary for a company of this stature, but a clause had to be added allowing for additional interest payments if BT's credit rating was downgraded further. Companies such as MobilCom, denied affordable access to either equity or bond markets, have only been able to raise bank loans while offering inflated yields because they are backed – though not necessarily indemnified – by larger operators with substantial shareholdings.

Not surprisingly, pressure has been exerted on equipment vendors to provide finance in return for the contracts to build new 3G networks. While this is normal practice, most vendors themselves issued profit warnings during 2001 and already have a lot of exposure to operators, and are therefore poorly placed to bail out 3G licensees – especially at less than commercial rates. Since vendor finance is normally passed over to financial intermediaries, the ease with which this can be done will anyway dictate how much new initial financing can be taken on by vendors. Great care must be taken to check the regulatory regimes governing licences in individual countries in relation to such matters as what happens if licences are in default or passed on, as well as tax implications. Nevertheless, for example, Hutchison 3G UK announced in April 2001 that it had secured total financing of $5.1 billion, of which $1.1 billion would be provided by Nokia, NEC and Siemens. Nokia also agreed to provide financing worth $1.8 billion to Orange.

Subsequently, in August 2001, Amena announced that it had secured a $2.1 billion loan as well as vendor financing from Ericsson and Siemens. Simultaneously, Hutchison invited bids for a 10-year $4.7 billion loan facility for H3G in Italy. Despite the fact that the lenders would have limited recourse to the parent firm – this nevertheless constituted the first occasion on which recourse to a parent company was provided – 15 banks tendered for the loan. In the end, the only way the money could be raised was via a $2.9 billion bank loan of which $1.9 billion had full recourse to Hutchison and the rest, while non-recourse, could only be drawn down subject to achieving performance targets; $0.9 billion of vendor financing; and $0.9 billion of either subordinated debt or shareholder equity.

It would appear that, by the end of June 2001, Nokia had provided a total of almost $5 billion in vendor finance, of which $4 billion had been refinanced in the markets. This represented roughly 17 per cent of the value of infrastructure sales, but presented no real difficulties for the cash-rich manufacturer. This largesse appeared to account in good part for the sharp upturn in Nokia's share of contracts from 11.3 per cent in the second half of 2000 to 33 per cent in the first half of 2001. During the same period, Ericsson's share fell from 43.6 per cent to 32.6 per cent after providing $2 billion in vendor finance with half being refinanced. Meanwhile Nortel, which had acquired a respectable market share of 11.3 per cent on the back of vendor finance provided to the likes of Amena in Spain, was unable to continue the policy during 2001 and as a result saw its market share fall by over one-half.[73]

Subsidies

The explosive growth of 2G was greatly assisted by the policy of handset subsidies in most major markets such as the UK. However, with markets approaching saturation, most 'new' customers already in possession of a handset due to churn, and new contracts increasingly taking the form of relatively unprofitable prepaid, these subsidies began to be withdrawn during 2001. The effects were fairly dramatic – Telefónica, for example, saw its subscriber acquisition costs drop by one-third during the first half of 2001 compared to one year earlier. Nevertheless, subsidies are expected to reappear when GPRS and 3G handsets become widely available. In particular, it is argued that if customers are to become used to paying out relatively large sums for sophisticated services, they must not be deterred by the prospect of paying out a large upfront sum for a handset. Furthermore, new entrants such as Hutchison are expected to build market share on the back of cheap, and possibly free, handsets, probably seeking to recover their subsidy by locking in the customer for more than one year.

Equipment sharing

As noted above, the issue of equipment sharing has particularly raised its head in countries such as Germany and the UK where licence fees are retrospectively seen as excessive. There has not yet been, and there is unlikely to be, a consistent regulatory response because of the differences among countries in licence conditions and attitude to health dangers. Furthermore, regulators are, unsurprisingly, tending to watch developments in other countries before making cast-iron rulings of their own. However, it can be argued that while the sharing of 'passive' infrastructure – sites, masts, antennas and power supplies – is likely to be relatively uncontroversial almost everywhere, the sharing of 'active' elements such as base stations and switching equipment, which offer much greater potential for savings – is another matter. Allowing for the greater complexity of shared facilities, the sorts of concessions mentioned under Germany above will, it is thought, permit a new entrant to save roughly 30 per cent of its anticipated capital expenditure on network roll-out although the majority of savings will appear towards the end of the build-out period. Other than new entrants, the obvious beneficiaries of sharing are operators seeking coverage in sparsely populated areas and operators in markets with more than four licensees.[74]

The European Commission has so far refused to take a firm stance on the issue, seemingly waiting for specific proposals to pass through the hands of member states' regulators before deciding whether intervention is required. It is certain that any dispute will end up with the Commission given that it is simultaneously trying to resolve the application of the concept of 'joint dominance' throughout the sector.

Although there are potentially sizeable sums to be saved, sharing will require operators to become interdependent in respect of issues such as speed of network build and quality – for example, will it be possible to use equipment from more than one vendor if they are separately contracted to the partners in the joint venture? – to a degree that most may cavil at when the time comes to sign formal agreements. Furthermore, disputes are likely to arise over sites – for example, about which should be used and under whose ownership – especially in densely populated areas where there is the possibility of signal interference. More fundamentally, perhaps, any savings may end up having to be passed on to consumers because smaller operators may try to use their savings to keep prices down in the hope of building up market share, and their larger brethren may be forced to respond in kind because service quality will differ very little as between operators. In addition, there may be consequent delays in consolidation which almost everyone now believes to be inevitable.

On the technical side, Nokia announced in May 2001 that its new multi-operator radio access network (RAN) would enable operators to share networks while retaining control over their own licensed spectrum, radio cells and services. Potential savings would amount to a maximum of 40 per cent of RAN outlays and operational expenditures. Availability is probably no earlier than the end of 2002.

Delays and debts

As noted, there is an apparent incentive for licensees to roll-out their 3G networks as fast as possible to recover any licence fees paid up-front. Certainly, during 2000/01 most operators faced a major cash drain not only because of licence fees but also because penetration was still rising sharply in many cases and they were investing heavily in 2.5G and Internet portals. However, in practice, most licencees now appear to be happy to delay as long as possible if only to give them an extended period to milk their 2G/2.5G networks. This willingness to delay is strengthened by the fact that many operators have withdrawn handset subsidies and ceased to chase marginal subscribers, which has had a dramatic effect on the amounts of cash flowing in rather than out.

The position for new entrants is somewhat different to that of incumbents because they cannot fall back on the 2G cash flow, but as we have seen, few new entrants pose much of a competitive threat and some may never build their 3G networks, so incumbents can afford to be sanguine about delays especially since they are taking cues about the speed of network roll-out from each other. Regulators may insist on the meeting of minimum rates of roll-out since these are technically subject to contractual conditions, but they may make concessions in practice and the penalties they can exact may not worry incumbents overmuch.

By delaying as long as possible, operators will be able to clean up their balance sheets and reduce their indebtedness, which is arguably their biggest short-term problem. For this reason, some of the fears about the horrendous implications for balance sheets of licence fees plus roll-out costs have probably been overstated.

Mobile virtual network operators (MVNOs)

We have already noted that MVNOs already exist in relation to 2G – there are nearly 50 of one kind or another in Western Europe – and that they are likely to be a feature on the 3G landscape. However, care is needed because there is no universally accepted definition of an MVNO.[75] A relatively rigorous view would require an MVNO to own a mobile switching centre, possess a unique mobile network code, issue its own branded SIM cards and operate a pricing structure significantly different from that of the owner of the network over which it provides its services. What it does not control are masts, base stations and frequencies. While a company such as Virgin Mobile appears to fit these requirements reasonably well, it is argued that it is strictly speaking tied too closely to One-2-One in the UK – in effect, it has formed a joint venture and the equipment is in reality all provided by One-2-One. Swedish-based Tele2 is arguably the best example of an MVNO in Western Europe because it does provide its own switches. As for most others, these are little more than basic resellers of other operators' services.

The central issue, unsurprisingly, is that once a prospective MVNO moves beyond the reseller model, it must start laying out appreciable sums for equipment installation and this leaves little latitude for decent profits. Since incumbents are aware of this, they take advantage by selecting virtual service providers to suit their own objectives and by keeping margins down to a minimum. It is possible to choose a middle path between true MVNO and reseller, along the lines of Tele1 Europe, by becoming an enhanced service provider – eschewing ownership of switches but possessing a billing, intelligent network and voicemail system, and a unified messaging platform.

Because a number of companies which either failed to obtain 3G licences, or diplomatically declined to bid, will want to become MVNOs, the licensees will have the upper hand in negotiations. However, this needs to be tempered by the knowledge that the spectrum was often so costly that it must be fully utilised as soon as possible, in which respect MVNOs may well prove useful. Certainly, some MVNOs, in the guise of enhanced service providers, will have marketing skills that are in short supply among traditional mobile network operators. What will happen when the latter rush out to acquire such skills, and hence come into direct competition with their related MVNOs, is another matter.

Some 3G licensing conditions expressly authorise the role of MVNOs without necessarily specifying what the latter are. However, as potential bidders in Hong Kong have pointed out, if MVNOs are permitted to bid for a share of the spectrum obtained by each licensee they could end up with more spectrum in aggregate than the licensees themselves. In the EU, licensing conditions have generally had nothing explicit to say about the role of MVNOs, although the forthcoming 'beauty contest' in Ireland provides additional spectrum for the winner of the 'A' licence if it agrees to allow an MVNO to use its network.

Roaming

As noted, the possibility that 2G will survive until as late as 2010 has raised the issue of inter-generational roaming. Such roaming can occur across a single network or between networks such that a new 3G operator without a 2G licence obtains access to an existing 2G network. To ensure that unfair advantage is not taken of this provision, it is normal to require that the 3G operator complete a specified proportion of the network before being granted inter-generational roaming rights. In addition, where operators form international joint ventures, these will need to incorporate roaming provisions, although the technical issues arising from international roaming between 2G and 2.5G networks are a more immediate concern. This is arguably just as well because when it comes to 3G roaming it is necessary not merely to allow for the different systems – for example, W-CDMA and cdma2000 – but also for different approaches to issues such as packet loss taken by the various equipment manufacturers. Furthermore, at the end of the day, despite the obvious difficulties of incorporating all kinds of roaming facilities in handsets, customers will doubtless want them to be as light as existing 2G handsets. A final factor is that in many cases charging for 2G services will be time-based whereas charging for 3G will be packet-based, which raises problems for billing, especially in relation to prepaid.[76]

Prospects for new entrants

Roaming also has a bearing on the prospects for new entrants. In cases where new entrants cannot quickly obtain a GSM or PCN licence, they become wholly dependent upon the right to roam onto an existing 2G network – not to mention the willingness of customers to churn subscriptions. Such a right may anyway only be conferred once the new entrant's 3G network has been partially rolled out – and hence after a good deal of up-front expenditure which is not compensated by any inflow of revenue – but in any event the crucial point is that whereas regulators generally mandate the right to roam, they may not necessarily want to mandate the charge to be levied for this service.

Self-evidently, most 2G incumbents, facing saturation in their markets, have no interest in fostering additional competitors, and hence can be expected to levy interconnection charges that leave little or no profit margin for the new entrant. In many cases, incumbents are anyway running short of spectrum to meet their own needs and feel that they should not be obliged to hand any over under such circumstances.

How this matter gets resolved then becomes a country-specific issue. For example, there is no mandatory roaming in Germany, but MobilCom is an established reseller and Group 3G (now Quam) has been able to connect up with the struggling E-Plus which has an interest in sharing the costs of rolling out its own 3G network. In contrast, the situation in France has been so problematic that, so far, only the two large 2G incumbents have obtained a 3G licence, and new entrants have not even bothered to bid – in part because they do not expect to reach an acceptable 2G roaming deal with either Orange or SFR.

The fact that Broadband Mobile has recently given up its licence voluntarily – admittedly the financial sacrifice was modest so this practice is unlikely to become commonplace – and that there have been no new entrants prepared to bid for a number of recent licences, suggests that the experience of the initial new entrants has put others off from following suit. Those that have already paid their licence fees have generally drawn in their horns. In Spain, for example, Xfera has decided to postpone unnecessary expenditure until its prospects have become clearer, although the various licensees involving Hutchison Whampoa are better financed and hence willing to push on for now. Those that persevere may eventually induce regulators to insist on lower interconnection prices, but are probably wise for now not to rely on it.

Those new entrants that survive are likely to end up in somewhat different hands to the original licensees as consortia members unwilling or unable to survive without an inflow of revenue sell out to those that are. Hence, it may be predicted that most new entrants will end up in the hands of operators with stakes in 2G incumbents in other member states.

Technical hitches

We have noted above that the earliest attempts to introduce 3G in the Isle of Man and Japan fell foul of various technical hitches, in particular relating to the issue of handing on signals from one base station to another. At this juncture it is difficult to be certain of the causal factors, but it is worthy of note that Qualcomm took advantage of the situation to claim that the synchronous base stations associated with cdma2000 would have avoided these problems, and that advocates of W-CDMA had brought them on themselves by adopting asynchronous base stations in order to avoid payments to Qualcomm for its intellectual property rights. However, its claim that W-CDMA could have used the GPS to synchronise base station clocks was met with scepticism given the ability of the US military to subvert GPS for its own short-term objectives.

Technical hitches are set to persist if only because the sophistication of 3G handsets dwarfs that of 2G predecessors. In general, 3G handsets will need big screens and a keyboard or stylus and handwriting recognition, or some form of alternative input technology other than voice. Batteries will need to last as long despite the much higher rate of drain,[77] and the whole package will need to weigh little more than 2G handsets. In addition, because of the initial need to roam over existing 2G networks to ensure full geographic coverage, 3G handsets will need to contain several chipsets. Finally, where cameras are introduced, these need to be kept away from antennas. And all of this has to be controlled by software which is glitch-free.

It seems that handsets may be modular in design to permit components such as the display, a camera, a keyboard and a speaker to be connected in a variety of configurations. Not surprisingly, the initial versions of 3G handsets did not even pretend to provide for all of the possibilities, and hence even once they become glitch-free this does not mean that new types of glitches will not reappear as handsets become more sophisticated.

At the end of the day, technological drawbacks notwithstanding, W-CDMA is likely to become the dominant technology, no matter what Qualcomm claims, for the simple reason that there are currently more than five times as many GSM subscribers potentially able to upgrade to W-CDMA as cdmaOne subscribers able to upgrade to cdma2000 (see the discussion on standards that follows). It may well be the case, therefore, that W-CDMA handsets will show a faster rate of development.

Standard wars

As noted previously, there are essentially three competing 3G standards – W-CDMA, cdma2000 and TD-SCDMA. Since the latter is initially going to be confined to China, assuming it is to be introduced at all, the standards war is in reality being fought between the first two. Of these, cdma2000 is generally considered to be technically superior, cheaper and faster to implement, but this is unlikely to prevent it losing the war.

The key reason is that W-CDMA is the chosen route for all those networks currently running GSM in both 900 MHz and 1800 MHz variants. This means, firstly, that it is the choice throughout Europe. That would be insufficient to win the day if other continents were to plump wholeheartedly for cdma2000, but, in practice, none has done so. The situation is that, for example:

- In Japan, DoCoMo and J-Phone are firmly in the W-CDMA camp with over 50 million 2G subscribers between them, and while KDDI has opted for cdma2000, its 'mere' 16 million 2G subscribers mean that it will almost certainly struggle to maintain market share.
- In South Korea, where there were 27 million CDMA subscribers at the time, all the bidders initially wanted to opt for W-CDMA. Although a consortium was formed to hold a cdma2000 licence, the technology was clearly seen as a second-best solution for companies unable to participate in W-CDMA. However, the situation has recently become rather confusing as all the licensees have opted to build a cdma2000 network first in order to to meet the deadline posed by the Soccer World Cup.

- The situation in China is ambiguous, not merely because of the uncertain status of TD-SCDMA but because the various GSM networks run by China Mobile, China Unicom and government agencies make China the world's largest market for GSM handsets. Hence, while Qualcomm is being allowed to make some initial inroads into China with cdmaOne in relation to Great Wall Telecommunications, this is largely a political move to assuage the USA related to China's application for WTO entry. It is also the case that China Unicom is rolling out a new CDMA network, but cdma2000 is unlikely to be anything other than a minority technology.
- In Brazil, the largest South American market, both technologies are likely to be introduced – for example, Telesp Celular has opted for cdma2000 1xRTT whereas Telemar has opted for GSM/GPRS. However, the fact that 1800 MHz licences are being auctioned covering a number of the cellular regions now means that national coverage via cdma2000 1xRTT is problematic and this may induce all operators eventually to switch to W-CDMA. Furthermore, the situation in Brazil precludes cdma2000 roaming throughout South America and provides some incentive for operators elsewhere to concentrate upon W-CDMA.
- cdma2000 has a fairly firm foothold in the USA where Qualcomm is based and the government prefers not to influence the choice of technology. For now, Verizon Wireless, Sprint PCS and the smaller Leap Wireless and NTELOS are in the cdma2000 camp, although Vodafone would much prefer Verizon Wireless to adopt W-CDMA since it intends to use it everywhere else, and AT&T Wireless, Cingular Wireless, Nextel and VoiceStream are in the W-CDMA camp with varying degrees of commitment. AT&T's (and partners') decision to cap investment in TDMA and adopt GSM/GPRS as an evolutionary pathway to W-CDMA, was especially damaging for cdma2000 prospects in the USA and Canada.

The issue of partners is of some importance, as noted in the Verizon Communications/Vodafone case, but it is also an aspect of the DoCoMo strategy of taking minority stakes in a variety of overseas telcos such as AT&T Wireless in order to put pressure upon them to adopt W-CDMA. It has also done so in the case of Hutchison Whampoa – it has a stake in Hutchison Telecom – which, as noted in Table 4.2, has an interest in a number of 3G licences both in Europe and the Far East.

It has been argued that switching technologies, as exemplified by AT&T, is uneconomic given the need to incur additional equipment costs. However, this view is disputed by Dundee Investment Research[78] which argues that evolving TDMA networks to 3G via EDGE would be even more expensive because EDGE would use up additional spectrum for data conveyance without increasing the capacity for voice telephony – of little significance in Europe given high penetration levels for 2G but important in the USA where lowish 2G penetration at 45 per cent means that there is scope for considerably more voice telephony. Furthermore, AT&T has managed to introduce GPRS much quicker than via an upgrade to EDGE.

Vendor alliances

It is too early to predict how the market for 3G equipment will be divided up among the various competing vendors, especially in relation to handsets which remain in a largely developmental stage for now. However, we can get some idea of progress from the network contracts to be found in Tables 3.1 and 4.1 and from the GPRS contracts in Table 2.1. We also have some specific information in relation to the market for 2G handsets from the likes of Gartner Dataquest which shows clearly that not only is Nokia the market leader but that it gained market share during 2001, reaching roughly a 35 per cent share of the annual global market of roughly 400 million handsets. A long way back in second place is Motorola with roughly 14 per cent, followed by Sony–Ericsson (SEMC – see below) and Siemens with perhaps 7.5 per cent. The main also-rans comprise Samsung of Korea, Panasonic (Matsushita), NEC, Mitsubishi, Kyocera, Alcatel and Philips.

Most analysts predict that market concentration will increase over time, driven in part by the switch to GPRS and 3G which present challenges beyond the resources of most individual manufacturers – and possibly all bar Nokia. The commonest prediction is that Samsung will overhaul SEMC during 2002 with Siemens losing ground. Given that most manufacturers had a torrid time during 2001, and that many were obliged to downsize in response, it is understandable that restructuring is taking place. This also reflects the fact that, whereas previously the Japanese manufacturers kept to their fast-growing home patch where GSM was not the exclusive technology as in Europe, the onset of GPRS followed by 3G – where Japan and Europe are introducing compatible technology – is giving them scope to muscle in on new markets at the expense of smaller European 'national champions' such as Alcatel and Philips.

However, the key issue for most manufacturers is how to compete against Nokia. Fortunately for them, Nokia is unlikely to expand its market share much further because this would leave network operators overly exposed to a single source of supply and they will want to keep at least a few other companies in business. Nevertheless, the sensible path, other than outright mergers which are for the moment unfashionable due to the heavily bombed-out share prices and depleted cash reserves of potential acquirers, is to form joint ventures/alliances. Here again, the sensible approach would seem to favour links between European and Far Eastern companies that could bring somewhat different strengths to the negotiating table and provide an entry into the main markets throughout the world with desirable modern designs.

It has to be recognised in this context that handsets are no longer seen as objects desired for their technical functionality alone. Handsets are increasingly seen as fashion statements and customers place considerable importance upon the availability of ring tones, games and other features. What Japanese manufacturers bring to the table is an unmatched understanding of consumer electronics devices needed for the development of colour screens and the means to support entertainment-based services, whereas European manufacturers bring their state-of-the art mobile technology and, crucially, relationships with the main network operators.

The history of joint ventures so far has been patchy at best. For example, in April 2000, it emerged that Alcatel and Lucent were holding talks with a view to a merger. The merger was allegedly driven by the need to gain economies of scale in the face of recessionary conditions rather than a need to offer a wider product portfolio. However, Lucent was determined to pursue a true merger of equals, whereas its collapsing share price and desperate need for cash led Alcatel to opt for an effective takeover with majority representation on the new company's board. At the end of May the talks predictably collapsed. Alcatel is now owner of a two-thirds stake in the Evolium joint venture formed with Fujitsu in May 2000.

Siemens and Toshiba announced an alliance to develop W-CDMA handsets in November 2000. The arrangement was that the vendors would jointly develop components but conduct their own marketing and sales. Although Siemens had a pre-existing infrastructure joint venture with NEC, this was not extended to cover handsets. In April 2001, Siemens disclosed that its handset division was losing money and that a 6–7 per cent market share was inadequate in the longer term. Although it was prospering in the Asia-Pacific region, the need for added strength in the USA led to rumours of talks with Motorola during the summer of 2001, but these were put on hold, such as they were, in October. In November, it was rumoured that Siemens and Toshiba would merge their mobile businesses, but in early December the alliance was disbanded.

Ericsson's market share dropped sharply from 10.9 to 7.5 per cent, comparing the first half of 2000 with that of 2001, as it suffered from design shortcomings and a failure to get its products to the market in good time. At the beginning of October 2001, Sony Ericsson Mobile Communications (SEMC) was created as a joint venture to which Sony's entire cell phone business was transferred. However, it was specified that SEMC would not take on board any of the parents' bad assets. SEMC set out to enter the US market where Sony had previously had a short-lived link with Qualcomm in 1999. SEMC's initial product is expected to be a GPRS handset towards the end of 2002.

SEMC followed on from an alliance formed in August 2001 between the two largest handset manufacturers in Japan, and arch rivals in the 2G market, NEC and Matsushita (Panasonic) which were contracted to supply 3G handsets to DoCoMo. It was claimed that neither would be able to develop a full range of 3G handsets independently although each would market under its own brand.

Whether these recent alliances will prosper is a moot point, but the widely accepted view is that unless vendors can achieve a 10 per cent market share – no problem for Nokia and Motorola; a possibility for Ericsson and, in the medium term, for Samsung; but a problem for all the others, and especially Alcatel and Philips whose market shares collapsed during 2001 – they will continue to lose money in ultimately intolerable amounts. If it survives in the short term, the Sony–Ericsson alliance will probably also be a long-term survivor, but, in general, the history of vendor alliances is not inspiring of confidence.

Health risks

As noted above, fears are increasingly being voiced about the dangers to health, especially in the form of cancers, arising from the ever-increasing use of mobile phones – more correctly from exposure to the radio-frequency (RF) fields generated by both handsets and their base stations. In 1996, the World Health Organization (WHO) set up the EMF Project to investigate and evaluate the scientific underpinning to claims about health risks, and the UK government established an independent expert group for much the same purpose. In some respects the problem has diminished in that modern digital handsets emit much less radiation than their analogue predecessors. Furthermore, base stations emit less radiation than handsets. Nevertheless, the topic has become front-page news, and fears expressed by the public, especially in relation to the erection of additional base stations in heavily-populated areas and near to schools, are genuinely felt even if difficult to support scientifically.[79]

Scientific studies into a link between RF fields and cancers are inconclusive, although there are individual cases of brain tumours that might possibly be attributed to RF fields. On that basis, a plethora of lawsuits has been launched in the USA on behalf of alleged brain cancer victims, accusing the industry of conspiracy, misrepresentation, negligence and product liability.[80] An obvious problem lies in estimating how much exposure there has actually been. Even if these cases are thrown out – none has made any real progress so far – it has to be recognised that since mobile handsets are used increasingly as a substitute for fixed-wire links, the risk may well be increasing even if attributable cancers have yet to become apparent.

The WHO and other bodies do not accept that a case has yet been made out to justify the use of RF-absorbing covers or other 'absorbing' devices on handsets. In any event, the effectiveness of such devices is uncertain. Despite this, a major strand in lawsuits consists of the argument that manufacturers took out patents between 1991 and 2000 for minimising radiation exposure or alerting users to high levels which it is alleged demonstrated that they were already aware of the radiation risks. In response, the manufacturers claim that the applications related only to the design, performance and efficiency of their products.

Even in the absence of provable links, restrictions are increasingly being introduced on the erection of masts on the grounds that they would cause nearby residents to worry.[81] This is likely to become a commonplace reaction, if only to protect planning authorities from a public backlash.

Pricing services

Given the paucity of operational services it is not possible to say anything much on this topic as yet. However, some information can be gleaned from the experience of DoCoMo's FOMA in Japan. The initial base price of a FOMA subscription has been set at $30 a month including $13 of voice or data calls, which is not a significant increase compared to i-mode. Data are priced per 128K downloaded and cost 60 per cent of the i-mode rate. However, 3G is designed to deal with much bigger data files, and if subscribers are prepared to pay $60 a month for their downloads, the cost falls to one-thirtieth of the i-mode rate. Hence, subscribers appear to be getting a superior service for much the same price, although handsets are inevitably more expensive. Further, because coverage is initially so limited it is probable that prices will fall as a mass market develops.

These prices are extremely reasonable given the high level of salaries typically found in the main Japanese cities. Since these are well above the average typically to be found in Europe, the monthly subscription charge would approximate to little more than $20 in terms of European salaries. This is a worrying small sum given that the cost of many European licences ran into billions of dollars, and it has to be remembered that the Japanese licences were free. Further, whereas the success of i-mode means that there is a natural progression for millions of Japanese subscribers to upgrade from 2.5G to 3G, the comparative failure of WAP and the late arrival of GPRS means that there may be a quantum leap for many European subscribers straight from 2G to 3G.

Regulation in the EU

It is first worth noting that, as the above discussion has made clear, the approach to 3G licensing within the EU has been somewhat inconsistent. This reflects the preference expressed by Member States for a fairly minimal level of harmonisation where licensing is concerned. Not surprisingly, the European Commission would prefer it to be otherwise, and is exerting pressure to this end. It has launched a proposed Framework Directive[82] whereby measures related to the use of radio spectrum envisaged by Member States will need to undergo a consultation process with both the relevant regulatory bodies in individual countries and the Commission. What the Commission wants is to be able to retain the ultimate power to prevent Member States from taking actions incompatible with the new framework which will harmonise the allocation, assignment and conditions of use of radio spectrum for all non-military purposes. It also wants to introduce a system of secondary trading in spectrum. It was never likely to achieve all of its aims since the European Parliament was opposed to any transfer of powers away from itself and the European Council, but in June 2001 EU ministers agreed that radio spectrum should be managed centrally by a committee including representatives of the Commission, Member States and carriers, while licences would continue to be allocated by individual governments.

Second, the Commission is concerned that the deployment of 3G networks in the EU will be adversely affected by such matters as the high licence fees paid in certain countries and the generally poor credit ratings of European telcos. Although it does not propose to deal with this via new legislation, it wants there to be a detailed dialogue within the EU addressing matters such as the possibility of delays in 3G network deployment; amendments to licence duration; the consequences of simultaneous roll-out requirements; conditions to be met in order to permit network infrastructure sharing; dealing with spectrum left over from the initial licensing round; the organisation of subsequent licensing rounds; and issues relating to the provision and acquisition of subsequent licensing rounds.

Conclusions

It may be noted that services accessible via a mobile handset are all available via a PC and hence they need to add value by taking advantage of factors such as a subscriber's known geographical position at any point in time. So far, efforts to achieve this in Europe have not been a success. For example, manufacturers such as Ericsson have not rushed to market with WAP phones in large volumes on the grounds that transmission speeds are still too slow to permit sufficient useful applications to be run. In any event, anyone used to graphics and animation is likely to be somewhat disappointed with WAP which has to strip them out to operate on a mobile handset.

The widespread introduction of GPRS will resolve at least part of the bandwidth problem, and at the same time will do away with existing restrictions on the richer content that has need to be removed, but precisely when is unclear.[83] Furthermore, while 2.5G is undoubtedly good enough to satisfy a significant segment of the market, especially households, if large companies are persuaded to buy 3G services then 2.5G network operators could find themselves left with the low-margin end of the market.

It is important to note that the overlap between ownership of PCs and mobile devices in the West has meant that Internet access via mobiles has been directed towards those already familiar with Web usage. In Asia, in contrast, Internet access via PCs is generally much lower, flat-rate unlimited access is not the norm and cable access is often unavailable. As a consequence, the i-mode approach has been phenomenally successful in Japan. In the many parts of the world where PCs are scarce and incomes are low, it may be extremely difficult to persuade people to use mobile for anything other than voice telephony.

It is also of significance that surfing is at best a somewhat tedious activity. Using a small mobile screen will accentuate that fact.[84] Hence, many believe that the critical breakthrough will be the development of voice-enabled access. Meanwhile, there is widespread agreement that the next few years will probably see the market split into two main segments: that needing huge amounts of bandwidth and hence 3G, and that needing personalised information that is time-sensitive and a facility to order, say, tickets as well as information relating to the location of the handset such as the nearest restaurant, which can be satisfied by 2.5G. However, the latter type of service is necessarily of low value – how much, after all, is it worth to know where to find the nearest restaurant?[85] At the end of 2000, a trial by Sonera – which announced that it could possibly manage without owning a 3G network by concentrating upon 2.5G via GPRS – admittedly using a small sample and using a combination of equipment that mimicked the likely capabilities of 3G handsets, indicated that news, specialised movie information and music was of interest, but not game playing.[86] Whether these represent 'killer applications' is another matter,[87] especially since a simultaneous study by consultancy OTR suggested, in a somewhat contradictory manner, that the only potential winners would be location services, games and sports and multimedia messaging.[88] The latter, an expanded variant of the SMS incorporating audio, graphics and pictures – see Chapter 1 – is viewed by Nokia[89] as the likeliest 'killer application' in the light of the unexpected success of SMS, and there is certainly much interest among consumers at the prospect of texting photographs. DoCoMo, in contrast, appears to believe that the most common use for a 3G handset will be to act as an electronic wallet, although the fact that users are able to see one another now that video phones with a tiny camera mounted on top have become available is likely to become extremely popular – unless, of course, the speaker is somewhere he/she is not supposed to be.

Just how 3G licensees will make a profit on their investment accordingly remains unclear. Certainly, they will need to rethink their relationships with their content suppliers and the methodology for sharing revenues with them. It must be borne in mind that 2G licensees make the bulk of their profits from selling simple voice and text messaging services on a massive scale and hence do not appear to be all that well placed to provide a sophisticated range of 3G data services. Furthermore, it remains unclear what prices any specific services will bear. Insofar that this puts consumers in the driving seat – able to dictate what services are worthy of payment – they have reason to be happy. On the other hand, it seems unlikely that anything by way of access or services can continue to be supplied without charge.

The ability to charge is, ultimately, the most fundamental issue, and on the face of it the experience of trying to charge over the Internet via computers has been a generally unhappy one. On the other hand, it can be argued that the mobile Internet offers much better prospects. This is because, according to Standage,[90] a mobile phone tends to be used by only one person and carried around by that person at most times, thereby permitting the network operator to identify both person and location; the network operator can set the default portal; and users are not merely accustomed to paying for data transmission in all forms but appear to accept – at least for now – that its transmission via a mobile device commands a premium price. The basic message – that content should often be supplied merely to generate transmission fees – is somewhat at odds with how most people expected 3G to be distinguished from 2G, but what 3G can obviously do is transport much more sophisticated content – for example, graphics are a splendid device for piling up data usage – with correspondingly high transmission fees. The i-mode experience (see Chapter 2) tends to bear this out.

Where the revenue side of the equation is concerned the situation is clearly therefore not all bad. However, it is not simply a case of providing additional services and sending a bill. Few doubt that large numbers of customers will eventually appear – possibly one billion by the end of the decade – if only because voice telephony will become cheaper with 3G. Nevertheless, most analysts do not believe that the figures add up. This is mainly because they almost universally assume that the rolling out of networks will be at least as costly in aggregate as the licences themselves – indeed, figures of up to $320 billion have been cited by the likes of Quotients Communications. In Germany, for example, this may not be true because of the high licence fees, but an alternative way of looking at the issue is to express the pay-back period as seven years or more where a 2G incumbent pays a nominal licence fee, and twice that where a sizeable fee is paid.[91] Not all licensees can be expected to survive such a long course, and hence the view is often expressed that there is only room in an individual country for three survivors at best – in the German case basically T-Mobile and the Vodafone Group and, perhaps, one other. Few see any realistic prospects for the new entrants which are stuck with the additional cost of building up a brand name while taking on well-established competition.[92]

A further cloud on the horizon relates to the claim by a Finnish professor that he has developed so-called Dynamics HUT Mobile IP software that will enable the launch of a 3G service at speeds of up to 11 Mbps in major cities in the UK earlier than the incumbents. Services will be supplied via Finnish ISP Jippii Technology using the 2.4 GHz spectrum that has yet to be licensed. Although hundreds of base stations will need to be installed on buildings, this technology is alleged to require much less investment than conventional 3G networks. However, the issue of licence fees has yet to be properly addressed because in some countries this spectrum is not meant to be used for commercial gain.

There is a parallel development in the USA where what is known clumsily as 802.11 technology – a wireless local area network (W-LAN) standard – uses the same spectrum, which is free and unlicensed, in its 802.11b variant sometimes known as Wi-Fi, for which there is a competing technical standard in the form of Bluetooth.[93] The cost of fixing a base-station transmitter with a range of 100 yards to a domestic building is only roughly $250, to which must be added $100 for a card to enable a laptop – a cheap way to access 11 Mbps and hence one which is being rapidly adopted in big cities and especially in airports, hotels and cafes as well as on school and university campuses. The crucial aspect of 802.11 is that it recognises that people on the move tend to become stationary before accessing the Internet. However, although there may already be one million users, it is a data-only system which cannot therefore supplant 3G but rather act as a complement to it. Furthermore, although suitable for laptops, it was originally designed to operate with desk-top computers and for now tends to require more power than can be handled by smaller battery-powered equipment such as PDAs and mobile handsets, especially in its 802.11a variant which will shortly also be available using spectrum in the 5 GHz range as will a competing technology known as HyperLAN2 used in Western Europe.

For the time being there are a variety of technical problems to overcome before 802.11's role in relation to 3G is clarified.[94] For example, individual locations in the USA tend to be under the control of specific ISPs, and the ultimate aim must be to link all of these together via a common billing system. Given the existing interoperability of equipment vendors' wireless systems, this is arguably as much an issue of organisation as of technology. Certainly, the individual pieces of a potentially widespread W-LAN network are being created well in advance of the roll-out of 3G. However, there appears to be a difference in approach between Western Europe and the USA since the dominance of mobile operators in Europe tends to favour their rolling up the development of W-LANs into the provision of GPRS/UMTS. Arguably, the development of W-LANs in Europe is being held back by the need to roll out 3G networks, but operators do appear to be aware that they cannot afford to cede control over W-LANs to other parties. Hence, for example, Sonera in Finland already has its wGate service in operation with the ambition of covering 200 'hotspots' by the end of 2002. Given the desire for integration, Sonera is basing its W-LANs on Nokia's GSM products which allows for the use of SIM cards for authentication. Whether a major mobile operator will seek to play the same role in the USA is as yet unclear.

It is possible to sum up the above in the following way. Some countries are – at least in terms of the conventional view of the development of 3G – still stuck at an early stage of 3G development. For example, some US operators have yet to switch to a fully digital network with a GPRS overlay. In most of Europe this stage has essentially been reached – although GPRS is not yet fully developed – and hence the main immediate issue is standardisation. This relates, for example, to platforms, with variants of i-mode trying to muscle in on WAP, and PDAs trying to muscle in on conventional 2G handsets. This brings with it the thorny issue of interoperability. When this has been sorted out, final decisions need to be made about the 'killer applications' – assuming such exist – and these must then be marketed on a sufficiently huge scale as to make the entire exercise economically viable.

As a final point it may be noted that the development of 4G is already under way – for example, DoCoMo announced a link-up with Hewlett-Packard in December 2000. As a preliminary step, 3.5G is forecast to be capable of delivering up to 10 Mbps in favourable conditions, perhaps as part of a wireless local loop system. The Japanese government, acting in conjunction with DoCoMo, KDDI, Japan Telecom, Sony, Matushita, NEC and Fujitsu, published an initial blueprint for 4G during May 2001 in the hope that it would be able to create a de facto protocol with a view to the introduction of services in 2010 – or even earlier if recent comments by DoCoMo are to be believed.[95] As from 2004, the telecoms ministry intends to begin reallocating spectrum already in use, with the possibility that compensation will be given for returned spectrum. However, European manufacturers such as Ericsson are also intent on having a major say in such a protocol. The forecast speed of 4G transmission is between ten and fifty times that of 3G, although the lowest figure is the most likely to be delivered.

A 10-year gestation period between stages in a technology is considered reasonable, but it will turn out to have taken considerably longer to switch from 2G to 3G if, as many analysts fear, services do not become widely available until 2004. There are several reasons why this should be so. First, standards are yet to be fully developed so handset manufacturers may produce equipment that will run with varying efficiency on different networks. Second, equipment may be slow to appear if only because so many countries are trying to achieve roll-out at the same time. Third, large numbers of new antennas are needed and, as indicated above, there is something of a public backlash against indiscriminate antenna construction which as expected gathered momentum during 2001. Fourth, the size of base stations partly depends on the amount of data to be handled, which remains somewhat hypothetical. Fifth, the handover of customers as they move from one 3G base station to another needs further work if the problem of lost connections is to be eradicated. Sixth, battery life is a problem because a 3G screen will have to be bigger and continuously backlit to cope with text and pictures, not to mention colour. Finally, a lot of work needs to be done on billing systems. Whether sufficient equipment can be installed, let alone tested, within the next two years accordingly seems to be increasingly implausible, notwithstanding deadlines laid down in licence conditions. A further factor is that for an appreciable period of time the 3G networks will be limited in geographic scope, and hence 3G handsets will need to have an in-built capacity to pick up signals from 2G base stations when the 3G signal expires. This means that handsets need to have another radio built in to 'hand-off' calls from the 2G to 3G sections of the phone.

Manufacturers will anyway want to clear inventories of older models before launching 3G handsets at competitive prices. In February 2001, Alcatel announced that it no longer expected 3G handsets to be widely available in Europe before 2004, and that it consequently expected GPRS to have a longer lifecycle to bridge the gap. It was also noted that its about-to-be-launched GPRS phone could only operate at 14.4 Kbps, barely above the speed of WAP although it did provide an 'always-on' facility, and that its only rival, made by Motorola, could only manage 19 Kbps. The financial markets, unsurprisingly, are clearly unenthused by events and the entire equipment manufacturing sector has been in what can best be described as turmoil.[96]

An educated guess, in the light of delays announced by DoCoMo and others, and Vodafone's September 2001 announcement about the expected initial speed of its European network, is that most urban dwellers can expect to be offered 3G at no more than 64 Kbps no earlier than late 2002, rising to 384 Kbps in 2005, while those living elsewhere may have to wait a very long time for anything beyond GPRS at perhaps 80 Kbps. When 2 Mbps will become available is anyone's guess, but quite possibly not before 2010.[97] However, as previously noted, W-LANs will provide the desired speed much earlier, albeit under limited circumstances, primarily for the business sector.

Consumers who are eagerly awaiting the high-speed music-playing and video-conferencing-capable handset with a built-in camera and a large colour screen beloved of marketing departments should accordingly scale back their expectations for now. Nevertheless, despite these words of caution, it is unwise to assume that the outstanding technical issues will not be resolved. Indeed, in February 2001, Nokia announced that its new mobile Internet technical architecture (MITA) framework, incorporating a multimedia messaging (MMS) platform called the Artuse MMS Center, would be capable of delivering rich content such as audio, still photographs and video clips. Its launch would be followed by a new m-commerce payment solution and an improved location service.[98] However, as noted previously, getting this technology installed on operators' 2.5G networks, let alone 3G networks, is another matter,[99] and negotiations between network operators and content providers are currently causing serious problems.[100] Ultimately, getting people to pay for the services provided, which consume a lot of bandwidth, is probably the trickiest matter of all.

6

A Case Study of Vodafone with a Piece of Orange and the DTs

A glance at Table 4.2 in the main text is sufficient to disclose that while there are a number of major players in the 3G market, two are apparently much more active than the others. As of January 2002, what is now the Vodafone Group had sixteen entries in the table, while what is now Orange had thirteen. However, the imbalance is in practice rather greater than it appears since, first, Vodafone is a major player in Japan where Orange is absent, and, second, Vodafone is a partner in Verizon Wireless in the USA where again Orange has no interest in a mobile operator. Equally, Table 1.3 indicates that, aside from Vodafone, the operator with the clearest global aspirations is Deutsche Telekom subsidiary T-Mobile which operates in the USA (via wholly-owned VoiceStream Communications) but not in Japan. For these reasons, the case study which follows concentrates upon Vodafone while assigning Orange and T-Mobile a secondary role. It is also significant to note that, like most of the other 3G players, Orange and T-Mobile are subsidiaries of a major fixed-wire operator, whereas Vodafone was created as, and has always remained, an almost exclusively mobile operator. The only other major 3G player that did not arise from the same origin is Hutchison, but in this case it invariably operates as a consortium partner and the ultimate parent, Hutchison Whampoa, has extensive interests outside the telecommunications sector.

Initial scenario

At the culmination of its takeover bid for Mannesmann, in March 2000, what is now the Vodafone Group was trading at almost £4 (roughly $6) per share. By June 2001, this had fallen to £1.40 (see Figure 6.1). On the face of it, this was nothing extraordinary by the standards of technology, media and telecommunications (TMT) stocks at the time, with many

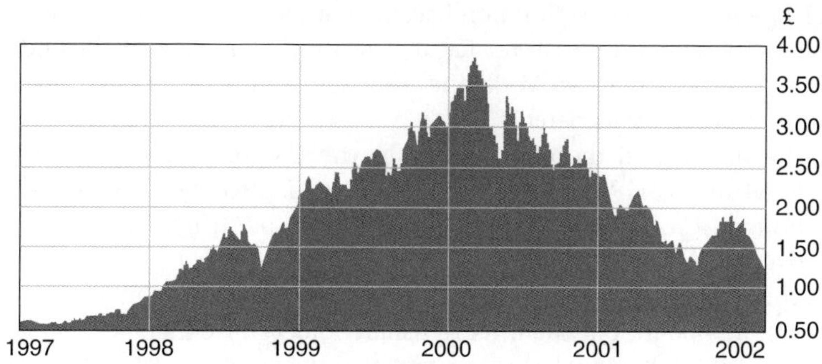

Figure 6.1 Vodafone Group – share price
Source: Daily price quotations in the Financial Times.

down by 80 per cent from their peak values in 2000, but the difference was that, immediately post-Mannesmann, Vodafone was one of the largest companies in the world, briefly worth over $300 billion and still valued at $230 billion at the end of 2000. In early May 2002, however, its market value temporarily fell below $100 billion. In other words, the shareholders in Vodafone had between them lost, on paper, over $200 billion, sufficient to purchase the entire manufacturing sector of most countries, and, indeed, quite a few complete countries. A dispassionate observer of these data would reasonably conclude that Vodafone must be one of the worst-managed companies in the world, but s/he would be wrong. Indeed, it was until recently viewed as one of the world's *best*-managed companies. To understand why a bit of history is in order.

Background

Prior to the takeover of AirTouch of the USA, Vodafone, somewhat unusually, was a fully listed UK-based company – that is, its shares had been floated on the stock market – and, interestingly, no other telecoms operator had acquired a significant stake. It had started out as a subsidiary of Racal, with an analogue licence dating from 1982, at which time it was the only private sector competitor to the then publicly-owned incumbent, British Telecommunications (BT).

Vodafone's initial success had a lot to do with the marketing skills of Chris Gent and his boss until 1997, Gerry Whent. The parent company knew about technology but not how to sell its products. Once Vodafone had generated considerable success in locking up the business market for mobile handsets, a flotation became both feasible and desirable as a means of generating cash for Racal. This duly took place at the end of the 1980s, and released Vodafone from the admittedly modest restrictions imposed by its parent.

The first digital mobile operator in the UK market with a licence granted in 1991, Vodafone became and remained the market leader, although it eventually faced competition from three other digital networks – Cellnet (half-owned by BT and licensed in July 1994), Orange (licensed in April 1994) and One-2-One (licensed in September 1993). BT bought out the minority shareholder in Cellnet, Securicor, in October 1999, having previously renamed the company BTCellnet. The controlling shareholder in Orange was Hutchison Whampoa, based in Hong Kong (see below). One-2-One was owned by MediaOne of the USA and Cable & Wireless of the UK. It was put up for sale in March 1999 and bought by Deutsche Telekom in August.

Vodafone was much the most internationally minded of the four UK mobile operators. By the end of 1998, it had acquired fairly extensive interests in Europe, some wholly-owned but usually in the form of minority stakes taken in consortia bidding for new licences. Elsewhere in the world it had, for example, acquired stakes in Australia, Fiji, Malta and South Africa. Nevertheless, it lacked a base in the two most significant markets outside Europe – the USA and Japan.

In considering its international strategy, Vodafone was able to take advantage of its unusual situation in that, unlike other UK operators and, indeed, most operators elsewhere, it was a pure mobile operation whereas the norm was a greater or lesser degree of fixed-mobile convergence. Furthermore, at precisely this point in time the financial markets were coming to appreciate the full potential of mobile telephony, and, hence were more than willing to back the expansion of the likes of Vodafone.

The crucial issue was that Vodafone would either have to expand so as to gather together a coherent worldwide network and gain economies of scale, or it could expect to be taken over by one of the European or American incumbents. Not surprisingly, it preferred the first option, and it was in a position to deliver because its rapidly rising share price meant that it could make all-paper takeover bids.

Vodafone/AirTouch

The various telecoms markets in the USA were undergoing massive structural changes in the aftermath of the Telecommunications Act of 1996. In January 1999, one of the Baby Bells, Bell Atlantic, made an all-paper takeover bid worth $45 billion for cellular operator AirTouch, but was poorly placed to raise its offer because its share price was relatively depressed at the time. AirTouch was already Bell Atlantic's co-partner in PrimeCo of the USA and in Omnitel of Italy. Vodafone, long considered as a potential partner for AirTouch and itself involved with it in joint ventures in Sweden (Europolitan) and Egypt, immediately made a counter-bid worth roughly $55 billion, consisting largely of stock together with a small amount of cash. This much higher offer was facilitated by the fact that Vodafone's own share price had soared during 1998. The offer was proposed as a merger of equals, but structured as an acquisition with a 50.1 per cent stake allocated to Vodafone's shareholders.

Aside from their existing collaborations, AirTouch also operated European joint ventures in Belgium, Germany, Italy, Poland, Portugal, Romania and Spain, whereas Vodafone operated in the UK, France, Germany, Greece, Malta and the Netherlands. Germany was, therefore, the only European country in which they competed at the time, and Vodafone was perfectly happy to dispose of its interest in E-Plus while taking on AirTouch's interest in Mannesmann Mobilfunk's D2 network. There was no overlap elsewhere in the world. This indicated that there would be few obvious cost savings, but collaboration in selling and technology was expected to pay dividends. Certainly, the two companies would together form the first more-or-less global mobile carrier, to be known as Vodafone AirTouch, operating on five continents. However, AirTouch was a regional operator in the USA and would have to take steps to acquire a national footprint to compete with the likes of AT&T Wireless and Sprint PCS.

Vodafone's offer was subsequently raised to $62 billion, comprising $85 in Vodafone stock per AirTouch share plus $7 in cash, at which point Bell Atlantic withdrew. By the end of June the takeover had been ratified by both sets of shareholders, the FCC and the European Commission – the latter making it conditional upon the disposal of Vodafone's interest in E-Plus.

In a fit of pique, reacting to the disappointment of a lost opportunity to stitch together a coast-to-coast network in the USA, Bell Atlantic had dissolved the PrimeCo arrangement with AirTouch in April. However, the synergies implicit in combining AirTouch's licences in 25 western states with Bell Atlantic's licences in 24 eastern states were potentially too good to miss out on, especially given the national reach of competitors AT&T and Sprint, and in September 1999 it was announced that the two companies were negotiating to merge their US cellular interests into a separate company (subsequently to be known as Verizon Wireless). The result was an arrangement whereby Bell Atlantic would end up with a 55 per cent controlling stake and the majority of board members. However, to compensate for its loss of control – regarded as worthwhile given that the alternative was to construct its own network from scratch at huge expense – and also by way of compensation for the premium that Vodafone had paid when bidding for the assets held by AirTouch, Vodafone AirTouch's 45 per cent stake would be larger than its contribution of 40 per cent of the overall number of subscribers in the new venture. In addition, Vodafone AirTouch would transfer $4.5 billion of debt to the venture. Finally, because Vodafone AirTouch was concerned that its minority interest would leave it in a vulnerable position should the relationship turn sour, it insisted that the venture should be partially floated after three years to provide it with an exit route.

Regulatory clearances were needed for the new venture, but the lack of significant overlaps and the existence of rival national networks were likely to lead to a positive outcome. Once its own proposed merger with GTE to form Verizon Communications was given regulatory clearance, Bell Atlantic would also be able to add GTE's mobile assets to the new venture at a later stage. Meanwhile, Vodafone AirTouch itself acquired further US mobile assets in the form of the (largely analogue) networks of CommNet Cellular, located in rural areas of nine western states. The cost of $764 million in cash plus the assumption of $600 million in debt was justified in terms of a reduction in the expense of 'roaming' agreements.

Mannesmann/Orange

Mannesmann had its origins in several industrial sectors – engineering, automotive and tubes. These continued to produce a good deal of revenue – over $15 billion[101] was forecast for 2000 – but the associated ebitda[102] was forecast at a miserly $300 million. It had diversified into telecoms and was known to be actively seeking to expand its telecoms interests in Europe. It had become involved in the ultimately successful bid by Olivetti for a majority stake in Telecom Italia, and was widely believed to have been the main beneficiary of that bid in that it had acquired stakes in mobile operator Omnitel and fixed-wire operator Infostrada at the very cheap price of $8.2 billion. It had also acquired the fixed-wire business of o-tel-o in Germany. These telecoms interests were expected to yield almost as much revenue as the industrial assets in 2000, but the forecast ebitda was close to $5 billion.

Under the circumstances, the Mannesmann chairman, Klaus Esser, who had been appointed in May 1999, was understandably keen to separate the company into two parts, each to be given its own listing. This process was instigated in September 1999. Mannesmann was expected to follow this up with a bid for UK mobile operator One-2-One, but surprisingly pulled out leaving Deutsche Telekom with a clear run. It subsequently emerged that it had already learned of the possibility that the Hutchison Whampoa stake in another UK mobile operator, Orange, would come on to the market in the fairly near future.

In mid-October 1999, it was revealed that Mannesmann was in talks with Hutchison concerning the future of Orange. It was not, however, a question of Hutchison wanting to bail out of its interests in wireless operators: what was proposed was for Hutchison to swap its large stake in Orange for a much smaller stake in Mannesmann. News of these discussions lifted the Mannesmann share price slightly to $162. When the cash plus paper deal was announced, it valued Orange at $31 billion, in addition to which Mannesmann agreed to take on Orange's debt of roughly $4 billion. For its part, Hutchison agreed to exchange its 44.8 per cent controlling stake in Orange for a 10.2 per cent stake in Mannesmann and to hold on to a minimum 8.5 per cent stake for 18 months.

The move on Orange, which was approved by the European Commission in December 1999 subject to the condition that Mannesmann sold Orange's stake in Connect Austria, sent a fairly explicit message to Vodafone AirTouch: You are the minority partner in D2 and Omnitel, and if we consider our interests to be best served through alliances with other potential partners then we will go ahead irrespective of your wishes.

Vodafone AirTouch/Mannesmann

This was not the message Vodafone wanted to hear. While the European mobile operations of itself and Mannesmann presented an excellent strategic fit, if Mannesmann acquired Orange then Vodafone AirTouch would be participating in several major joint ventures with a co-partner that appeared to be as much of a competitor as an ally.

It did not take very long for the financial markets to conclude that the only rational response for Vodafone AirTouch would be to make a takeover bid for the whole of Mannesmann. Indeed, France Télécom immediately informed Vodafone AirTouch that it would be happy to take Orange off its hands were such a bid to succeed. What was certain was that Vodafone AirTouch would not be able to forestall the Mannesmann offer for Orange since Mannesmann did not need to seek shareholders' permission to issue the requisite number of additional shares and Hutchison had given an irrevocable acceptance in response to Mannesmann's offer.

Before launching such a bid Vodafone AirTouch had to take account of a long list of complicating factors. For example, Mannesmann had a core of loyal shareholders, no hostile takeover had ever been successful in Germany, and trade union representatives filled half the seats on Mannesmann's supervisory board. Since they could expect Vodafone AirTouch to dispose of the industrial assets that provided most of the jobs at Mannesmann, with probable ensuing job losses, they would inevitably oppose the bid. These were by no means trivial problems, so why was Vodafone AirTouch so keen to proceed? In the first place, its existing network in Europe consisted of too many minority stakes, and the addition of the stakes held by Mannesmann would greatly strengthen its hand, especially in the major economies. Secondly, there was the lure of sheer size. On a pro-rata basis, and adding in the wireless assets of Bell Atlantic, the merged company would have roughly 42 million subscribers. Not only would this reduce the need for roaming agreements with other carriers, but more of the latter would have to pay for Vodafone AirTouch to deliver their calls. Further, equipment manufacturers would have to pay increased attention to the specific technical requirements of the merged company, especially in the run-up to the introduction of third generation systems. Finally, the investment in developing a platform for mobile data and Internet services would be much easier to justify if the resultant technology was to be adopted on a much larger scale.

Hence, despite these difficulties, Vodafone AirTouch decided to press ahead with a 'friendly' bid valued at 43.7 of its own shares for each Mannesmann share – equivalent to roughly €203 per share or just over €100 billion (roughly $90 billion – one Euro was typically worth $0.9 at the time) for the entire company. This offer was dismissed by Klaus Esser as 'wholly inadequate'. The reply by Vodafone AirTouch was to 'go hostile' on 15 November. Klaus Esser responded with a vigorous defence. In September, he had announced that Mannesmann would be splitting off its automotive and engineering divisions into a separate listed company, with its partial flotation pencilled in for Spring 2001. He now announced that this would take place much earlier, and that interested parties were already in preliminary talks to acquire stakes. He added that Mannesmann had received acceptances from 75 per cent of Orange shareholders and that it was now Europe's largest mobile service provider. He also sought to emphasise the virtues of an approach based upon the integration of fixed-wire and mobile networks, rather than the pure mobile approach advocated by Vodafone AirTouch. Chris Gent, chief executive of Vodafone AirTouch, responded that he did not intend to dispose of any fixed-wire assets acquired as part of a takeover of Mannesmann. He also stated that he would be retaining the whole of the Mannesmann workforce and that any existing plans to hive off non-telecoms assets would be allowed to proceed.

One of the interesting by-products of the takeover of Orange by Mannesmann was that the latter's foreign ownership had expanded to 65 per cent – mainly because Hutchison Whampoa had swapped its holding in Orange for a 10 per cent stake in Mannesmann. In addition, UK institutions held 25 per cent and US institutions 25 per cent. Nevertheless, Vodafone AirTouch faced an uphill struggle to acquire even a simple majority of Mannnesmann shares unless it persuaded at least some German shareholders to sell out.

At the end of November the financial markets began to take the view that Mr Esser's defence was insufficiently persuasive, and that the odds were improving that the bid would succeed. Mr Esser responded like the true Anglo-Saxon he was not, stating that he believed in shareholder democracy and that he would concede defeat if Vodafone AirTouch secured over 50 per cent of Mannesmann shares. However, he indicated that a fair take-out price would be at least €300, effectively leaving Mannesmann shareholders with a majority stake in the merged company. The formal offer was tabled on Xmas Eve with Vodafone AirTouch announcing that the bid was no longer conditional upon clearance by the European Commission. The closing date of the offer was set at 7 February 2000.

On 19 January, Chris Gent appeared to make some concessions. He stated that he was prepared to consider an increase in the offer price, albeit in Vodafone AirTouch shares rather than in cash, but only a very modest one and that under no circumstances would it result in lifting Mannesmann's share of the merged company to more than 48.9 per cent as against the original 47.2 per cent. This, he claimed, was the best offer his own shareholders would tolerate. He also stated that he would consider holding on to Mannesmann's fixed-wire operations.

Despite his refusal to cede majority control to Mannesmann there was no sign of a 'white knight' riding to the latter's rescue. At the end of the day, the cost was simply too high for rival bidders to pay. The biggest potential problem therefore seemed to be the attitude of the European Commission, which confirmed that it had begun its first-stage inquiry with a deadline of 17 February. At the very end of January, Vivendi of France created despondency in the German camp by announcing that, far from wishing to form an alliance with Mannesmann, it had agreed to join forces with Vodafone AirTouch. This 50/50 alliance between VivendiNet, a joint venture between Vivendi and Canal Plus, and Vodafone AirTouch was temporarily named Multi Access Portal (MAP), and was being set up to distribute content and services to the 70 million (assuming Mannesmann was taken over) TV, computer and mobile phone screens connected to the networks of the two parties. In return for its support, Vivendi secured an assurance that Vodafone AirTouch would sell it half of Mannesmann's stake in Cégétel in France, almost certainly at well below its true market value, thereby raising its own stake to 51.5 per cent. On the face of it, this was a poor deal for Vodafone AirTouch, especially since there was no certainty that Vivendi would have struck a deal with Mannesmann, but it did not ultimately appear to be too big a sacrifice in the greater scheme of things. As in respect of the arrangement with Bell Atlantic noted above, Chris Gent appeared unusually ready to make pragmatic concessions in the pursuit of longer-term objectives.

The financial markets now took the view that the bid would succeed, boosting Vodafone AirTouch's share price on successive days to end 2 February at £3.85. This meant that each Mannesmann share was valued at €344 and the overall bid at €178 billion. The following morning it was confirmed that Mannesmann had accepted the bid on the basis of 58.96 Vodafone AirTouch shares per Mannesmann share, worth €353 at that time, giving Mannesmann shareholders 49.5 per cent of the new company. In June, the European Commission cleared the takeover subject to the commitment to sell off Orange and the provision of access to other operators of roaming facilities and wholesale services for a period of three years.

Orange

In mid-April 2000 the European Commission announced that it would not pursue a second stage investigation but would authorise the takeover subject to fairly modest conditions already accepted by Vodafone. Somewhat surprisingly, it did not insist that Orange be sold off. It also required Vodafone to open its network to other operators over a three-year period at prices identical to those charged by the network to Vodafone. Hans Snook, the Orange chief executive, declared his opposition to a sale even though potential buyers, allegedly including France Télécom – held to be extremely anxious to take remedial action in the light of its failure to acquire E-Plus (see below) – KPN, MCI WorldCom and NTT DoCoMo, were queueing at the door.

The future of Orange was finally resolved at the end of May 2000 when an offer was accepted from France Télécom which consisted of roughly $19 billion in cash together with 129.2 million of its shares worth roughly

$15 billion and the assumption of debts of $2.4 billion and of future liabilities of $5.5 billion relating to the UMTS licence in the UK. This worked out at an enterprise value of roughly $7000 per subscriber which was accepted as reasonable by the financial markets. An unexpected feature of the deal was that it was organised as, in effect, a reverse takeover of France Télécom's mobile operations via a company to be named Orange France, with the new organization run by Orange management, using the Orange brand and listed under the Orange name in London.

In order to avoid the Vodafone Group ending up with an undesirably large stake in the French operator, replacement loan notes would be issued and a complex set of call and put options issued which would serve to reduce the stake gradually over time. The European Commission stated that it was unlikely that there would be any need for a lengthy investigation given the previous discussions with the various parties, and the deal was cleared in mid-August subject to the disposal of the Orange stake in KPN Orange in the Netherlands, enforced because of the existing holding in Mobistar. Also, because Mannesmann received shares equivalent to more than a 10 per cent stake in France Télécom, the European Commission insisted that this stake be reduced to 9.9 per cent through a repurchase by the French telco. Mannesmann was granted neither voting rights nor a seat on the board.

Orange was thereafter in a position to compete head-to-head with the Vodafone Group in most European Union markets, in respect of both 2G and 3G, as shown in Table 6.1.

While Orange, with Hans Snook remaining as CEO, had embarked upon some minor acquisitions for cash during the summer as part of its global Internet strategy OrangeWorld, it was recognised that a partial flotation, initially pencilled in as 15 per cent, would be needed in order to make acquisitions using shares in Orange, which would have rolled up into it France Télécom Mobiles (FTM). In September, Orange made its first foray into South East Asia via the $142 million purchase of a 37 per cent stake in Bitco, a holding company with a 97.25 per cent stake in Thai mobile operator Wireless Communications Service (WCS) which was about to roll out a PCN network. Currently, Orange holds a 48.9 per cent in CP Orange Company, the renamed WCS. Other interests include 67.8 per cent of MobilRom in Romania, 64.0 per cent of Globtel in Slovakia (to be rebranded as Orange), 51 per cent of Vista Cellular in Botswana, 70 per cent of Société Camerounaise de Mobile, 86 per cent of France Télécom Dominicana, 85 per cent of Société Ivoirienne de Mobiles, 33.6 per cent of Société Malgache de Mobiles and 26.0 per cent of BPL Mobile in India.

Table 6.1 Vodafone Group v. Orange: 2G mobile assets in Western Europe (per cent holding) 20/01/02

Country	**Vodafone group**	**%**	**(subscribers 000s)**	**Orange**[1]	**%**
Austria	*			ONE*	17.5
Belgium	Proximus*	25.0	(1 014)	Mobistar*	50.7
Denmark				Orange*[6]	53.6
France	SFR*	31.9	(3 699)	Orange*[7]	100.0
Germany	D2 Vodafone*	99.6	(21 824)	*[8]	
Greece	Vodafone*[2]	52.8	(1 523)	Vodafone*[2]	3.0
Ireland	Eircell	100.0	(1 701)		
Italy	Omnitel Vodafone*	76.5	(13 335)	Wind*[9]	26.5
Netherlands	Vodafone*[3]	70.0	(2 262)	Dutchtone*[10]	100.0
Portugal	Vodafone*[4]	50.9	(1 419)	Optimus*	20.0
Spain	Vodafone*[5]	91.6	(6 995)		
Sweden	Europolitan*	71.1	(788)	*[11]	
Switzerland	Mobile Com*	25.0	(878)	Orange*	87.5
UK	Vodafone*	100.0	(13 164)	Orange*	100.0

Notes

* UMTS licence holders, sometimes as part of a consortium.

1 The holding company, Orange France is now 86% owned by France Télécom and 14% by private shareholders.

2 Formerly Panafon (rebranded January 2002).

3 Formerly Libertel (rebranded in January 2002).

4 Formerly Telecel (rebranded in October 2001).

5 Formerly Airtel (rebranded in October 2001).

6 Formerly Mobilix (renamed May 2001).

7 Formerly Itinéris (renamed June 2001).

8 Via its 28.5 per cent stake in MobilCom.

9 This fell to 26.5 per cent subsequent upon ratification of the acquisition of Infostrada by Enel on behalf of its subsidiary Wind at the end of 2001. There is an option to increase the stake to 32 per cent once Wind is part-floated.

10 France Télécom's 50 per cent stake in KPN Orange was sold to KPN for €500 million in cash in December 2000. KPN will be able to retain the Orange brand until 1 October 2002 after which it will be returned to Orange. Prior to this, the KPN network will be rebranded as 'Base'.

11 Via its 95 per cent stake in Orange Sverige.

Source: Compiled by author.

In November, Hans Snook stood down as CEO causing analysts to downgrade their estimates of what Orange was worth. When analysts ran their valuation models in December 2000, they concluded that Orange was now worth only $27 billion at best, and that with FTM rolled up into it the enlarged company would be worth at best $67 billion compared to estimates of $125 billion only six months previously.

The value range announced in the run up to the flotation was established as $50–$58 billion, with 633 million (13 per cent) of the total shares – increased to 4853 million through a new share issue of 50 million shares forming part of a 2 per cent allocation to directors and current/retired employees – to be floated at an institutional price of between €11.5 and €13.5 apiece. A further 95 million shares would be made available subject to demand (the 'greenshoe' allocation), and retail customers in France, Germany, Italy and the UK would get a €0.5 discount per share. In addition, two-year exchangeable bonds equivalent to a maximum of 243 million shares (including a 15 per cent 'greenshoe' allocation) would be sold at a premium of between 22 and 27 per cent to the institutional issue price and yield between 2.25 and 2.75 per cent.

Anxious to preserve its single-A credit rating in the face of heavy 3G expenditures, France Télécom announced that Orange would not be making any significant acquisitions for the time being. Its balance sheet would look strong because it was intended to transfer to it only $5.4 billion in debt, but it was losing money outside France and the UK and difficulties were anticipated in rebranding the various pieces under the Orange banner and standardising their technologies.

In early February, having traded below the minimum price for the IPO in the grey market, the price range for the institutional offer was lowered to €9.5 to €11, potentially reducing the value of Orange by a further $9 billion. It was felt by analysts that Orange should trade at a discount of roughly 10 per cent to Vodafone, the share price of which had fallen sharply. In the event, the institutional price was fixed at €10 – a mere 1 per cent discount to Vodafone – valuing the company at €48 billion – not that much more than was paid for Orange alone. Despite this, at the end of the first day's trading the price had fallen to €9.4 (see Figure 6.2 where the share price is converted to pounds sterling). The value of the floated shares was $5.7 billion, with a further $800 million potentially to come from the greenshoe operation. Shares were bought by 1.33 million individuals, obtaining full allocations if applying for fewer than 675 shares, 75 per cent of the excess up to a maximum of 8000 shares and 6168 shares if they applied for over 8000. The convertible bond was fully taken up, raising $2.4 billion with potentially a further $360 million to come from the greenshoe operation.

During the second day of trading the Orange share price fell again to €8.81, dragging down its parent to an 18-month low at €72. Altogether, a remarkable 60 per cent of the shares changed hands during these two days. Towards the end of March the share price finally rose well above the €9 barrier, then through €10 in early April, and €11

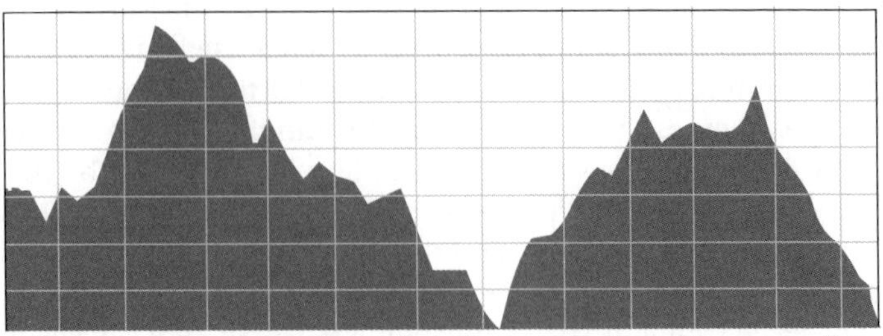

Figure 6.2 Orange – share price
Source: Daily price quotations in the Financial Times.

in mid-April, but subsequently fluctuated considerably along with other TMT shares, falling below €7.50 in September.

The rebranding of networks is continuing. In June 2001, for example, the Itinéris brand used in France was relaunched as Orange in the hope that it would represent a less staid image. It may be noted that it is fairly unusual for a market leader to be rebranded, and even more so for it to adopt the tag-line, in this case 'The future's bright ... the future's Orange', established in another country. In some ways, the future did look reasonably bright, with Orange increasing its subscriber numbers to 11.9 million in the UK at the end of June 2001, thereby becoming the second-largest UK operator, and to 35.5 million worldwide, although its net debt was on the high side at €6 billion. It announced its intention to operate in 50 countries by the end of 2005.

In October 2001, Orange increased its stake in Orange Sverige to 85 per cent by buying out Bredband Mobil's stake, leaving the other shareholders as Skanska (10 per cent), NTL (3 per cent) and Schibsted (2 per cent), and subsequently also acquired the Skanska stake. It denied that it intended to sell a further tranche of equity either to reduce its debts or to acquire E-Plus. However, a major problem was looming in respect of MobilCom in Germany, in which Orange held a 28.5 per cent stake. MobilCom had acquired a 3G licence as MobilCom Multimedia, and had debts of $10 billion. Orange wanted it to launch its network under the Orange brand as quickly as possible to get its brand name established, but MobilCom's largest shareholder, Gerhard Schmid, wanted to be bought out and appeared to be trying to force a dispute over strategy that would have triggered such a purchase – one that both

Orange and its parent were desperate to avoid because of the cost of the stake plus the need for one of them to take MobilCom's debt onto its balance sheet. By threatening both to delay the roll-out as long as possible and to launch under the MobilCom brand, Mr Schmid was clearly pushing Orange into a corner in the largest single European market.

As noted, put options were created when Orange was sold by Vodafone, and these have recently come home to roost. In March 2002, for example, France Télécom will be obliged to repurchase 50 million shares at €100, several times the ruling market price. Admittedly, the huge capital loss this will incur will not appear on Orange's balance sheet, but as a result of another put option granted to E.ON in November 2000 when it bought E.ON's stake in Orange Communications of Switzerland, France Télécom will be forced to repurchase 2.1 per cent of Orange shares outstanding at €9.25, which will undoubtedly exceed the market value at the time.

E-Plus

A further issue to be considered is the future of E-Plus, the third-largest mobile operator in Germany with over 7.5 million subscribers (out of a total of roughly 56 million at the end of 2001). As noted above, Vodafone AirTouch already knew that it would obliged by the European Commission to dispose of its holding in E-Plus because it was expecting to acquire Mannesmann's interest in the larger D2 network in Germany. Since they were themselves going through a restructuring, and wished to sell off peripheral interests, Veba and RWE of Germany – trading as VR Telecommunications – also decided to dispose of their own interests in E-Plus at the same time. As a result, 77.5 per cent of E-Plus became available – at least in principle – in early December.

France Télécom, licking its wounds after the destruction of its relationship with Deutsche Telekom in the wake of the latter's attempt to go it alone in an ultimately doomed attempt to take over Olivetti, was extremely anxious to acquire the Vodafone AirTouch stake in E-Plus, and negotiated a deal to buy the available shares for €9.1 billion. However, it had not bargained for the fact that BellSouth of the USA, the owner of the residual 22.5 per cent stake, would exercise its preemption right to buy the stake at the same price. This price was a bargain because the take-over bid for Mannesmann had triggered a rapid rise in the value of all European mobile interests, and BellSouth promptly made a deal to sell on the stake to KPN of the Netherlands at a profit. BellSouth agreed to retain its own 22.5 per cent holding for a further 18 months, after which point it would either continue to retain the stake, or convert it into either a 19 per cent stake in KPN or a €6.4 billion stake in KPN's mobile subsidiary. The purchase of the Vodafone AirTouch stake by BellSouth was cleared by the European Commission at the end of January 2000. The purchase of the residual VR Telecommunications stake by BellSouth followed on shortly afterwards and was cleared by the European Commission in February 2000 as was the taking of joint control of E-Plus by KPN and BellSouth – with KPN buying out its partner in January 2002.

T-Mobile

It is significant that the state has remained the largest shareholder in Deutsche Telekom (DT) with a 45.7 per cent stake. DT recognised fairly early on that a strategy of demerging would make sense, commencing with a small stake in its internet subsidiary T-Online. However, it has had no success so far with an IPO of its mobile subsidiary, T-Mobile.

Like other European incumbents, DT has had a fairly torrid time over the past year or so. The most obvious reason for DT's financial woes is its profligate expenditure. In particular, DT spent a substantial sum of money on 3G licences – $7.7 billion in Germany, $6.1 billion in the UK and another half billion dollars or so in Austria, the Netherlands and Poland. Roll-out costs will possibly come to as much again. The financial markets now doubt that these licences will break even, let alone earn a profit. As noted, DT had to watch impotently as Vodafone swept into Germany with the acquisition of Mannesmann. True, DT's own purchase of Vodafone rival One-2-One in the UK was revenge of a kind, but it ran up more debt, partly related to the need for a 3G licence, and left DT's collection of overseas assets looking patchy at best, with the only other major stake in Western Europe its wholly-owned Austrian subsidiary Max.mobil. The sale of its Sprint FON stake has left it without a fixed-wire network in the USA, and it has no significant assets in Japan, nor, indeed, in most of the Asia-Pacific region.

At the end of July 2001, the results for the first half of 2001 revealed a rise in net indebtedness to just under $60 billion due to the VoiceStream and Powertel purchases. Deutsche Telekom announced that the planned flotation of T-Mobile would be postponed until 2002 – a decision effectively underpinned by the agreement of the main credit agencies not to downgrade as a result – and that its mobile internet division T-Motion would become the first in Europe to charge for content (aping DoCoMo in Japan). Although Deutsche Telekom's coffers were topped up with the disposal of 50 million shares in Sprint PCS at roughly $25 apiece in

August 2001, the expectation was that, unless remedial action was to be taken as soon as possible, its net debt would rise above $65 billion.

It intends to cut this back via an IPO of roughly 20 per cent of T-Mobile sometime in 2002, raising perhaps $10 billion. However, despite the large number of T-Mobile's controlled subscribers, it has to be recognised that over 40 million of these come from Germany, the UK and the USA alone, that T-Mobile will remain a majority-owned subsidiary of a highly-indebted and significantly state-owned incumbent, and that establishing a company-wide brand – T-Mobile – which is currently recognised only in Germany, present major obstacles for the future.

Rebranding

Mobile service users in developed markets are a fairly fickle group, seemingly willing to switch brands without much concern as to whether the new service is likely to be much different to that currently consumed – which in general it is not. It is as yet unclear whether the mobile market can ever develop strong brand loyalty, but operators understandably prefer to believe in the possibility. For this reason, it was announced in July 2000 that in order to establish a global brand name Vodafone AirTouch would henceforth become the Vodafone Group and trade, almost everywhere, simply as Vodafone – this name will be used for convenience in what follows.

One obvious disadvantage for Vodafone compared to Orange is that the latter has a compelling set of brand values and a resonant advertising image that has helped to generate rapid subscriber growth combined with relatively low churn. Not surprisingly, therefore, France Télécom is moving rapidly to make Orange its universal mobile brand. By comparison, the Vodafone brand can be said to lack any kind of 'personality'. It can be argued that Vodafone is much the largest and most successful mobile operator, so this does not appear to have held it back overmuch. Nevertheless, having collected a set of assets that are being marketed under a variety of brands in different markets, Vodafone is in need of a global brand image although there is clearly a downside to replacing a recognised and successful brand name in certain of its markets.

The intention is that the Vodafone brand will eventually become universal, but the initial stage was to combine it with an existing brand with the latter dominant – for example, the ex-Mannesmann brand D2 in Germany became D2 Vodafone, in Italy it traded as Omnitel Vodafone and in Portugal as Vodafone Telecel. Similar rebranding also occurred in Greece, the Netherlands, Spain and Sweden, but as yet not in Belgium and France because Vodafone is not the controlling shareholder in the relevant operator. Vodafone already trades under its own name in Malta and Hungary, but cannot rebrand in Poland and Romania for the same reason. The obvious time to switch exclusively to the Vodafone brand will be with the launch of UMTS services, but the process has been accelerated somewhat in practice with Vodafone becoming established as the brand in Portugal and Spain in October 2001 and in Greece in January 2002. Germany followed in March and Sweden in April.

Tidying up

In February 2000, Vodafone AirTouch offered to buy all outstanding Mannesmann convertible bonds at the rate of 1440 new Vodafone shares per convertible bond. Just how much value that represented was becoming a moot point as the latter's share price went through a period of considerable volatility that was also affecting other companies in the telecoms sector. The various issues that needed to be resolved are discussed in what follows.

Airtel: In April 1999, Grupo Endesa and Unión Fenosa, opting to concentrate their interests in Retevisión and needing to avoid conflicts of interest as required by the regulator, had announced that each would put its 8.14 per cent stake in Airtel for sale. Banco Santander Central Hispano (BSCH) decided to exercise its right of first option to buy the shares, thereby raising its stake to 30.45 per cent, but it also announced its intention to sell the shares on to the other shareholders when the price had risen sufficiently.

Airtel remained one of the few European mobile operators without a controlling shareholder. Hence, in December 1999, seeking to take advantage of the opportunity thus presented, the now enlarged Vodafone offered to exchange the BSCH stake for roughly 5.5 per cent of its own shares, worth €7.2 billion on the day of the offer, seeking thereby to gain majority control. However, just as the deal was about to be signed, BT not only stepped in with a higher paper bid, equivalent, on the day of the offer, to roughly a 5 per cent stake in BT, but also invoked a shareholder agreement binding the companies that had seen off the bid by Retevisión which specified that any additional acquisition of Airtel equity by any of them would, if the others so wished, have to be shared out equally among them.

In January 2000, Vodafone signed an agreement with the three minority partners in the shareholding pact by way of a 'put' option giving it the exclusive right to buy their combined 16.9 per cent stake should they decide to sell, but the ownership issue still required final resolution. The compromise eventually reached left both Vodafone and BT as shareholders, but with the former firmly in control with a majority stake. Initially, BSCH sold its stake to Vodafone, receiving in return a stake of over 3 per cent in the latter and becoming its largest shareholder. Vodafone then built up its stake to 73.8 per cent overall by further purchases from other shareholders. The residual shareholding was divided between BT (17.8 per cent), Acciona (5.4 per cent) and Torreal (3 per cent). Vodafone then paid $1.6 billion in cash for the BT stake in May 2001 and dropped plans for a flotation. Airtel was rebranded as Vodafone in October 2001.

Atecs: It had been expected that the sale of Atecs, the engineering and automotive unit of Mannesmann, would be pushed through by the Mannesmann supervisory board before Chris Gent obtained the all-clear from the Commission and officially took over. Thyssen Krupp had tabled a €8.75 billion bid, but this was trumped by a €9.1 billion joint bid from Siemens and Robert Bosch (which was subsequently increased to €9.6 consisting of €3.1 payable by September 2000, €3.7 payable by the end of December and €2.8 of assumed pension liabilities). In the event, the offer was accepted shortly after Vodafone took over. In August, the European Commission authorised, first, the sale of Mannesmann subsidiary Demag to Siemens and Bosch, and, secondly, the sale of three engineering subsidiaries, Dematic, VDO and Sachs, to Siemens. However, it opened an investigation into the sale of Rexroth to Bosch which was eventually authorised in December subject to Bosch selling its existing piston pump business.

In July, Vodafone sold its ex-Mannesmann watches business to Richemont of Switzerland for €1.6 billion. This left as a non-telecoms interest only the 8 per cent stake in Ruhrgas. A further Mannesmann-related disposal took place in November 2000 when Warburg Pincus acquired Mannesmann iPulsys, the Dutch managed IP service, for an undisclosed sum.

Celtel Uganda: In August 2000, Vodafone sold its 36.8 per cent stake in Ugandan mobile operator Celtel Uganda to Netherlands-based Mobile Systems International Cellular (MSI)

China Mobile (HK): In October 2000, Vodafone signed a MoU to buy a 2.18 per cent stake in China Mobile (Hong Kong) for $2.5 billion in cash, and formalized the contract in mid-January 2001. It has first refusal to increase its stake should the opportunity arise.

Eircell: In December 2000, Vodafone made a successful all-paper bid worth €4.5 billion, including €250 million debt taken over, for Eircom of Ireland's mobile subsidiary Eircell. The offer consisted of 0.478 Vodafone shares for every two Eircell shares. A new subsidiary Eircell 2000 was created for the holding. Eircom promised not to create a replacement mobile operator during the ensuing three years except in certain 'limited circumstances'.

Infostrada: In October 2000, Infostrada was provisionally sold to Enel of Italy (see also below). The antitrust authorities stipulated that Enel could proceed only if it sold 5500 megawatts of generating capacity representing a profit reduction of over $1 billion a year. As a result, the sale price was renegotiated, with the offer falling in value by 34 per cent to €7.25 billion plus €1.3 billion of debt, but payable in full in cash rather than partly in Enel bonds.

Iusacell: In January 2001, Vodafone agreed to pay $973 million in cash for 34.5 per cent stake in the second-largest mobile operator in Mexico, Grupo Iusacell. The profitable operator was already part-owned by Vodafone partner Verizon Communications, holder of a 37.5 per cent stake.

Japan Telecom: As of the beginning of December 2000, Japan Telecom, one of the main rivals to NTT and, like it, a 3G licensee, was part-owned by AT&T and BT via a holding company whereby the former held a 10 per cent stake, partly via AT&T Wireless, and the latter 20 per cent, with the voting rights split 15 per cent apiece. In turn, Japan Telecom held a 54 per cent stake in J-Phone Communications (JPC), a holding company for its mobile interests, which, with nine million subscribers, was a major rival to NTT DoCoMo. Vodafone also held a 26 per cent stake in JPC with the remaining 20 per cent owned by BT. In turn, J-Phone Communications controlled the three operating companies, J-Phone East (JPE), J-Phone West (JPW) and J-Phone Tokai (JPT). Japan Telecom held a 51.2 per cent stake and Vodafone a 8.3 per cent stake in JPE, with equivalent stakes of 50.6 per cent and 10.1 per cent in JPW. BT's interests in JPE and JPW were much smaller at 1.2 per cent each and, unlike the others, it had no stake in JPT. Japan Telecom also operated the J-Sky service providing Internet access for four million subscribers.

This complicated situation meant that BT and Vodafone would be keen to battle it out for control of the real jewel in the crown – JPC. During December, Vodafone bought a 15 per cent stake in Japan

Telecom from two of its founding railway companies for $2.2 billion in cash. Also in December, responding to a planned 16 per cent stake in AT&T Wireless to be taken by DoCoMo, AT&T put its Japan Telecom stake up for sale. BT arguably had a pre-emptive right to buy the stake via its existing arrangement with AT&T, but Vodafone was also very anxious to buy the stake and had a much bigger cash pile to hand. In late-February 2001, Vodafone offered to buy the shares for $1.35 billion in cash, a 37 per cent premium to the ruling market price, provided AT&T could deliver them by the end of April. With this purchase, Vodafone ended up with a combined 39.5 per cent direct and indirect stake in JPC. This grew further in May 2001 when cash-strapped BT agreed to sell its various interests in Japan Telecom for $5.3 billion. Vodafone now owned a 45 per cent direct stake in Japan Telecom, a 46 per cent direct stake in J-Phone, and enlarged minority stakes in the three J-Phone operating subsidiaries. It also inherited from BT options to buy a further 5 per cent stake in the J-Phone subsidiaries.

In August, Vodafone arranged for J-Phone's four operating subsidiaries to be merged under the J-Phone Co banner in November. As a result, Vodafone ended up with a 39.7 per cent stake in J-Phone while Japan Telecom ended up with 45.1 per cent, but Vodafone's stake in Japan Telecom meant that its economic interest in J-Phone would rise to almost 60 per cent. In October this became larger again, at 69.7 per cent, when Vodafone bought from East Japan Railway Co roughly an 8 per cent stake in Japan Telecom for cash as part of a successful tender offer to buy 21.7 per cent overall for $2.6 billion – the share price had fallen by 57 per cent between mid-July and mid-September and Vodafone offered a 53 per cent premium – and thereby gained control of Japan Telecom with a 66.7 per cent stake. It is not intended to sell off the loss-making fixed-wire business for the time being nor to list J-Phone.

Mannesmann Arcor: Arcor is the only fixed-wire operator in Germany other than Deutsche Telekom to serve both business and residential customers with a local loop. In late March 2000, Vodafone announced that the need to raise funds to bid for UMTS licences was proving problematic, and that it accordingly intended to float a stake in the newly acquired fixed-wire operators, commencing with 25 per cent of the former Mannesmann Arcor which was worth roughly €4 billion at the time despite being loss-making. However, although this had been pencilled in for March 2001, it was indefinitely postponed in December 2000.

The central problem lay with Deutsche Bahn, the railway company, which held a 18.8 per cent stake in Arcor but a blocking 25 per cent plus one share of the voting rights – the other shares were divided between Vodafone with 73.2 per cent and Deutsche Bank with 8 per cent. Deutsche Bahn's first concern was that the trains were controlled by a telematics system built into the Arcor network, and it wanted to be sure that the system would remain safe. Secondly, it was under pressure to clean up its balance sheet and was reluctant to sell its stake at the bottom end of the market. Arcor is proceeding for now with further purchases of city carriers providing local loops such as ISIS in Dusseldorf and WuCom in Wurtzburg – members of BREKO, a non-commercial alliance of roughly 30 city carriers – which were acquired prior to the Vodafone takeover. In December 2000, it accordingly also acquired a 51 per cent stake in Netcom Kassel for an undisclosed sum.

It was subsequently anticipated that Arcor, which introduced a flat-rate Internet access offer in April 2001, would engage in an IPO sometime in late 2001. This is now expected to be delayed to 2003 although an earlier trade sale is likely. Meanwhile, Arcor DB Telematik will be created at the beginning of 2002, 50.1 per cent owned by Arcor and 49.9 per cent by Deutsche Bahn, to take over Arcor's role of telematics provider to the railways, comprising roughly 25 per cent of its turnover. Deutsche Bahn has the option to take full control of the new company from April 2005. Arcor will also transfer all of its rail-related tangible assets to Deutsche Bahn in return for roughly $1 billion in cash, with the latter retaining an 18 per cent stake in Arcor.

Ruhrgas: Vodafone inherited an 8.2 per cent stake in Ruhrgas, held via a 23.6 per cent stake in Bergemann, as part of the purchase of Mannesmann. This was sold on to E.ON in October 2001 for roughly $750 million in cash.

Shinsegi: In August 2001, Vodafone sold its 11.68 per cent stake in Shinsegi Telecom for an estimated $150 million and withdrew from the South Korean market.

Swisscom Mobile: In November 2000, it was announced that Vodafone would be buying a 25 per cent stake in Swisscom Mobile in March 2001 assuming that the latter had been partly floated off from its parent and had been successful in acquiring a 3G licence in November 2000 – which it was. The purchase from Swisscom would form part of a strategic partnership with Vodafone paying in cash, shares or both at its discretion, with the price of the equity element to be fixed as an average of the price ruling during the final few days before closing. In addition, an enhanced service provider agreement would be signed between Mannesmann

Mobilfunk and Swisscom's German subsidiary Debitel, authorising Debitel to access Mannesmann's UMTS services. In March 2001, Swisscom shareholders authorised the sale at a price of $2.64 billion – approximately half on closing the deal and the rest plus interest no more than one year later with $1.4 billion paid in cash at the end of September.

Tele-ring: Austrian fixed-wire and PCN operator tele-ring was previously majority owned by Mannesmann. In November 2000, it acquired a 3G licence in Austria trading as Mannesmann 3G Mobilfunk, subsequent to which Vodafone bought out the other stakeholders, Osterreichische Bundesbahnen, the Verbund and Citykom Austria. Despite this, it sold its entire holding to Western Wireless International of the USA for an undisclosed sum in May 2001.

Verizon Wireless: In April 2000, the situation in the USA partially resolved itself with the authorisation of the takeover of GTE by Bell Atlantic to form Verizon Communications. This made it possible to proceed with the previous agreement between Vodafone and Bell Atlantic to create their joint venture, named Verizon Wireless, with Verizon Communications as majority shareholder with a controlling interest. It was announced that Verizon Wireless, now with 24 million subscribers, would seek a stock market listing which would value it at in excess of $100 billion. However, adverse market conditions delayed the listing. It was widely believed that Vodafone would try to take control of Verizon Wireless, if only because it wanted to use the W-CDMA version of 3G to which it was committed elsewhere whereas Verizon Communications preferred the incompatible cdma2000. So far this has not transpired.

Vizzavi: In June 2000, MAP was launched under the renamed Vizzavi banner to provide e-mail, search facilities, information, entertainment and e-commerce. It will be the default home page for the three companies involved and utilise mobile phone, PC, TV and personal digital assistant platforms. Content will primarily be provided by VivendiNet (Canal Plus, Havas and Havas Interactive). It will operate initially in France, Germany, Italy and the UK with the existing Vodafone and Vivendi operating companies being rolled up into Vizzavi and offered a 20 per cent stake in the national subsidiaries. Vizzavi was cleared by the European Commission in July subject to the requirement that customers should be able to access third-party portals, change the default portal for themselves or authorise a third-party portal to alter the default portal on their behalf.

Despite the authorisation, Vizzavi became mired in a dispute with BT which took Vizzavi's owners to court in France, alleging that Vivendi and Vodafone should have involved their partners in Cégétel under the terms of the latter's shareholding pact – the ownership structure at the time consisted of Vivendi (44 per cent), BT (26 per cent), Vodafone (15 per cent) and SBC (15 per cent). The claim was dismissed, but the door was left open for further action relating to Cégétel mobile subsidiary SFR, owned 80 per cent by Cégétel and 20 per cent by Vodafone. As it happened, the court also gave Vodafone permission to transfer half of its 15 per cent stake in SFR to Vivendi, leaving the latter with a 51.5 per cent controlling stake although it would not be able to take direct control until the shareholding pact expires on 24 September 2002. By the Spring of 2001 it was widely believed that both BT and SBC would welcome the sale of their stakes, and Vivendi moved to prevent these falling into Vodafone's hands by claiming a pre-emption right should the shares become available. Nevertheless, with Vivendi increasingly interested in resolving the fall-out from its acquisition of Canal Plus and Seagram, and the formation of Vivendi Universal, it seemed probable that Vodafone would eventually emerge with control of Cégétel and SFR. As a first step, Vodafone agreed in June to swap its 15 per cent stake in Cégétel for 12 per cent of Vivendi's stake in SFR – on the face of it a technicality but in line with Vodafone's desire to withdraw from stakes in fixed-wire operators.

In April 2001, Vizzavi announced that it had acquired 700 000 registered users, that its UK portal had launched a location-based service in April and that it would seek a separate listing by the end of 2003. In addition to the existing Vizzavi services in the Netherlands and the UK, Vivendi Universal would transfer its online operations in France in May 2001 and set up Vizzavi sites in Italy and Germany by the year-end. By December the number of customers had risen to six million, but revenue remained modest with $45 million predicted for 2002 – a poor return on an investment of roughly $1.5 billion.

Vodafone Australia: Part of the Vodafone Pacific group (also incorporating New Zealand and Fiji), Vodafone Australia was frustrated in its attempt to merge with Cable & Wireless Optus. As a consequence, increasing competitive pressure forced it to announce redundancies in May 2001.

Doing the sums

It was possible to do some sums relating to Vodafone's financial situation as of June 2000. Its spending commitments were understandably huge even though the actual purchase of Mannesmann was entirely financed with its own shares. In the first place, it paid $9 billion for its UK UMTS licence, and that excluded the roll-out costs of perhaps $3 billion. It felt obliged to bid in the other main markets such as Germany where auctions were due to take place in 2000–01, even if as part of a consortium, so an approximate overall cost was $32 billion for licences and $14 billion for roll-out. However, it also needed to acquire spectrum in the USA where its existing network badly needed upgrading, and this could cost an additional $32 billion overall. Elsewhere in the world costs were expected to be more modest, perhaps $7 billion in total. Naturally, these were no more than informed guesses, so the overall total of $85 billion was subject to possible bounds of, say, $80–90 billion, but at the end of the day the sums involved would be very large.

So how to pay? So far, as noted above, the sale of parts of Atecs had raised $9 billion, and the sale of Orange had produced roughly $19 billion in cash. Its stake in France's Cégétel was being transferred to Vivendi for roughly $7 billion, and, despite its earlier protestations, stakes in fixed-wire interests such as Infostrada and Mannnesmann Arcor were on the block, although they were thought to be worth no more than $3.2 billion. The money expected to flow from the partial flotation of Vodafone Pacific had to be pushed back into the future with a nominal expectation of $7 billion. In the even more distant future, the partial flotation of the new US joint venture, Verizon Wireless, was expected to generate much bigger sums, but no decision had as yet been made about how much would be sold and when.

So there appeared to be approaching $40 billion already guaranteed with the potential for considerably more in the future. Bearing in mind that many commitments on the payments side of the ledger were also somewhat distant, the situation looked to be under control for the time being, and any shortfalls relatively easy to manage via the issuance of bonds and other instruments. Vodafone itself claimed that it would have net debt of $12.5 billion in March 2001.

In practice, Vodafone achieved a better result than expected when it agreed to sell Infostrada to Enel for $10 billion plus $1 billion of inherited debt outstanding at the end of June 2000. $5 billion was to be paid in cash, $2.2 billion in one-year bonds and $2.7 billion in three-year bonds. A further maximum payment of $0.5 billion would depend upon Infostrada's performance over the subsequent nine-month period.

Analysts felt that, partly for this reason, Vodafone's own estimate of its financial situation was misleading. It was estimated that once Arcor and the equity stake in France Télécom was sold, Vodafone would emerge with a positive cash pile worth perhaps $18 billion – which would make it stand out compared to its often heavily-indebted peers.

This did not, however, help the hugely deflated Vodafone share price to recover. After the bid for Mannesmann was successfully concluded in February the share price suffered a reversal, but soon recovered to reach a peak of almost £4 in March. Subsequently, the share price began to fall inexorably, eventually bottoming out at £2.40 at the end of May – a drop of 40 per cent from its peak. This reflected a general loss of faith in the prospects of telcos, but in particular serious reservations about the cost of acquiring UMTS licences. Between March and October the share price broadly fluctuated between £2.60 and £3.20, but mid-October saw another severe dose of bearish sentiment about the prospects of all telcos, driving the Vodafone share price to a new low of £2.35. It traded in the £2.40–2.60 range during most of the ensuing three months, but the beginning of 2001 saw the share price under pressure once again reflecting the share overhangs (see below).

The financial difficulties were compounded by other factors. In Germany, for example, the purchase of Mannesmann stimulated Deutsche Telekom subsidiary T-Mobil to cut tariffs and raise handset subsidies. Vodafone responded in kind, with the result that customer acquisition costs rose to $90 a head, with repeat costs whenever replacement handsets were purchased. Subscriber numbers duly soared, improving the longer-term outlook but simultaneously damaging short-term profitability.

A further issue currently hanging over the share price is the stake held by Hutchison Whampoa which it acquired when Mannesmann was bought by Vodafone. In March 2000, roughly 1.5 per cent was sold for $5 billion via the world's largest share placing at a 7 per cent discount to the market price, and a further 0.9 per cent was sold via a convertible bond in September at £2.80. The three-year bond carried a yield of 2.875 per cent and was convertible at a 27 per cent premium to the price ruling at the time of issue. Hutchison was left with a 3.47 per cent stake, and was not keen to hold on to it once the lock-up period agreed as part of the Mannesmann takeover had expired because it had a stake in an UMTS competitor to Vodafone in the UK. It promised not to make further sales without consulting Vodafone, and needed to take account of the fact that the market price was unattractive at the beginning of 2001. What transpired in January 2001 was an announcement that it would issue a $2.5 billion convertible three-year bond equivalent to roughly a 1.1 per cent stake in Vodafone, in the process depressing Vodafone's share price to a 12-month low of £2.15 compared to the bond's strike price of £3.10.

This was not, however, the only potential source of overhang since over 5.5 per cent of Vodafone was transferred during the purchase of its enlarged stake in Airtel and more was transferred in return for its stakes in Swisscom, Eircell and Japan Telecom (see above).

A final consideration is that the purchase of Orange by France Télécom, agreed when the latter's share price stood at €143, involved a put option (right to sell) linked to a floor value for the share price which the agreed formula delivered at €106. In practice, having repurchased 15.4 million shares at the original issue price in return for loan notes, France Télécom was committed to repurchasing the remaining 113.8 million of its shares held by Vodafone in three instalments during 2001/02, commencing with 58.2 million at €104.2 in early March 2001 and followed by 5.92 million at €99.66 at the end of March and 49.73 million at a minimum of €100 at the end of March 2002.

Table 6.2 Other Vodafone Group holdings, December 2001

Country	% holding	Main partners	Proportionate customers
Albania (Vodafone)	76.9	Panafon-Vodafone	41 000
Australia (Vodafone)	95.5	MCHL	2 120 000
China (CMHK)	2.2	n/a	1 134 000
Egypt (Click GSM)	60.0	EFG Hermes, Alkan, MSI	84 000
Fiji (Vodafone)	49.0	Telecom Fiji	36 000
Hungary (Vodafone)	50.1	Antenna Hungaria, RWE	169 000
India (RPG Cellular)	20.6	RPG Cellular	14 000
Japan (J-Phone)	69.7[1]	Japan Telecom	6 726 000
Kenya (Safaricom)	40.0	Telkom Kenya	285 000
Malta (Vodafone)	100.0[2]	Maltacom[2]	105 000
Mexico (Iusacell)	34.5	Verizon Communications	590 000
New Zealand (Vodafone)	100.0	n/a	1 044 000
Poland (Plus GSM)	19.6	TDK, KGHM, PKN, PSE	518 000
Romania (Connex)	20.1	TIW, ROMGSM	294 000
South Africa (Vodacom)	31.5	Telkom, Rembrandt	1 609 000
USA (Verizon Wireless)	44.1	Verizon Communications	12 658 000
Globalstar	7.2		
Arcor	74.0		
Vizzavi	50.0		

Notes

[1] As of November 2001, including the indirect holding via a 66.7 per cent stake in parent Japan Telecom.

[2] Subsequent upon acquisition of the stake held by Maltacom.

Source: Adapted from www.Vodafone.com/worldwide.

Vodafone seems happy to leave investors somewhat in the dark about its finances. In mid-November 2000, for example, it emphasised the proportionate ebitda for the first half of 2000, which revealed a surplus cash flow of roughly \$5 billion. In contrast, the conventional accounts revealed a a pre-tax loss of roughly \$6 billion because Vodafone had paid much more than net asset value when buying AirTouch and Mannesmann and needed to write this off as goodwill. It was true that such a write-off did not involve a cash outflow as such, but the accumulated debts of roughly \$20 billion in September needed to be serviced, and the Mannesmann purchase meant that earnings per share would fall sharply in the short term.

In late-February 2001, with its share price trading at £1.85 because of the delay to the Infostrada sale discussed previously, Vodafone's capitalisation fell below that of BP Amoco – and to not much more than it paid for Mannesmann – and it ceased temporarily to be the largest company in Europe. However, on a more optimistic note, Vodafone was able to announce in March 2001 that it had reduced its borrowings to \$10 billion following the sale of Infostrada. By comparison with almost all other telcos at the time this was an extremely modest figure, but it was insufficient to restore Vodafone's share price above the £2 mark. Curiously, Vodafone also sought to boost its share price by no longer counting inactive customers who had not made a call during the previous three months – almost 10 per cent of the 83 million worldwide total. The benefits would arise from the fact that average revenue per user (ARPU), which was apparently in decline because of the trend towards pre-pay packages, would thereby be enhanced. It went on to calm market fears about indebtedness by promising to concentrate upon customer retention rather than acquisition; not to go on an acquisition spree; and to introduce GPRS as part of a more gradual roll-out than originally expected of its third-generation networks.

The agreement to buy BT's stakes in Japan Telecom, J-Phone and its operating companies in early May caused a reappraisal of Vodafone's prospects to take place. On the face of it, Vodafone's net debts at the time stood at not much more than \$10 billion allowing for the cash paid for the AT&T stake in Japan Telecom. However, the BT stakes were set to add roughly \$5.3 billion directly to this total, and there were two additional factors which potentially arose because Vodafone would end up with a combined direct plus indirect stake of over 50 per cent in J-Phone. First, Vodafone would have to find the majority of the remaining cash – perhaps \$6 billion – needed to finance J-Phone's new 3G network. Secondly, as majority owner, Vodafone could be obliged to consolidate J-Phone's existing debts of $10 billion onto its own balance sheet. Vodafone argued that this would not be necessary because it did not have a 'dominant influence' over J-Phone. The simultaneous purchase of the 17.8 per cent BT stake in Airtel initially added a further $1.6 billion to Vodafone's debts, but Vodafone moved immediately to place a block of new shares at £1.94, the total value of which was raised from $4.3 billion to $5 billion – representing 2.8 per cent of the issued share capital – due to bouyant investor demand, thereby covering the bulk of its immediate outgoings for the various purchases. The fact that the new shares could be issued at a very small discount (10 per cent) to the ruling market price was a positive sign, although little immediate upturn could be anticipated given the pre-existing share overhang and, indeed, those subscribing to the issue very soon had cause to regret their behaviour as the share price began to plummet.

Revising the strategy and redoing the sums

In May 2001, Vodafone reported better than expected results for the year ending 31 March 2001. Ebitda rose sharply to $10.5 billion, with half coming from continental Europe and one-quarter from the USA. Pre-tax profit rose by 90 per cent to $6 billion due to the various purchases during the year, although write-offs relating to Mannesmann meant that a fairly meaningless overall loss of $12 billion was reported. Net debt stood at less than $10 billion, a mere 5.4 per cent of capitalisation. The $15 billion anticipated cost of rolling-out 3G networks was budgeted to come from its own resources. The main worry was that Vodafone was not consolidating its share of the debts of associated companies such as J-Phone and Verizon Wireless, leading analysts to estimate true indebtedness at roughly $23 billion – hence the need for the share placing discussed above – but broadly the overall picture was much healthier than that of its rivals despite Vodafone's admission that roughly 10 per cent of its subscribers were inactive (in which respect it was probably quite typical).

Nevertheless, by mid-June the Vodafone share price had fallen to £1.70 as the share overhangs began to weigh heavily on the market. KPN and Telia were now entitled to sell 370 million shares acquired via the sale of stakes in Eircell – admittedly less than a typical day's turnover – and Airtel shareholders were about to become entitled to dispose of up to $5 billion of Vodafone shares.[103] Further, Vodafone moved to mop up the holdings of 7,400 small shareholders left over from the Mannesmann purchase with an offer worth $500 million in total.

As viewed by analysts, Vodafone needed to address three central, inter-related issues: to integrate its recent acquisitions; to switch from a strategy involving the pursuit of all-out growth in subscriber numbers and market share to one emphasising the improvement in profit margins and cash-flow growth; and to bring 3G to profitable fruition. Central to this would be the withdrawal of handset subsidies (e.g. for pre-pay handsets in the UK), the creation of common products and services (e.g. a pan-European roaming tariff for both contract and pre-pay subscribers, and including both voice and text messaging), and the creation of a common brand, albeit tacked on at first to the known brands that Vodafone had aquired (e.g. D2 in Germany). It would also be possible to squeeze suppliers' margins given Vodafone's enhanced buying power, but the introduction of GPRS and 3G would probably require additional, hopefully short-term, handset subsidies. Any growth would occur via exploiting existing geographic concentrations, especially with a view to increasing ARPU which has tended to move inversely to subscriber numbers, although opportunities to increase potentially strategic stakes, especially in the likes of China Mobile (HK), would be taken up.

According to analysts' discounted cash flow models in May 2001, Vodafone shares should have been trading at roughly £2.70, indicating significant upside potential once the share overhangs were cleared, and, if 3G did prove to be a success, Vodafone was expected to be the major beneficiary. The first of the overhangs were duly cleared in June when Telia sold 80 millon shares and KPN disposed of 220 million at £1.56, in the process setting off a further decline to almost £1.40. The apparent decision by BSCH to hold on to its 1.84 billion shares then set off a modest upturn in the share price, but it subsequently announced that it had cut its Vodafone stake from 2.71 to 1.62 per cent, stopping the share price in its tracks. In any event, analysts had by then largely given up on TMT stocks for the year and were talking in terms of stabilisation around the £1.50 mark, representing a monumental loss of value since the Mannesmann takeover.

During the three months to the end of June 2001, Vodafone's proportionate subscriber numbers grew by over three million and a further 7.1 million were added due to acquisitions, raising the total to 93.1 million. However, the share price stayed down and, indeed, fell below £1.40 during July when Vodafone announced that it was struggling to obtain GPRS handsets and hence the introduction of its services in Europe would be delayed, albeit hopefully not until Xmas; that 3G handsets would be slow to arrive in bulk, and hence that it would be delaying the full launch of 3G services until 2003 in the UK, while saving money in the interim by reducing the number of base stations to be built during 2001 from 1250 to 750; that ARPU had fallen in both the UK and Germany; and that it had heavily written down the number of active users in the UK to 10.5 million, leaving it in third place behind Orange and BT Cellnet.

In late September, Vodafone paid $1.4 billion in cash as the second tranche of its payment for its 25 per cent stake in Swisscom Mobile, but recovered roughly half this sum the following month by selling off its stake in Ruhrgas inherited with Mannesmann. Figures for the third quarter indicated pre-paid subscriptions running at 88 per cent across the entire company with some signs that markets were stabilising. However, this was insufficient to forestall the announcement of 600 redundancies among UK-based staff in mid-October.

In November, Vodafone announced its half-year results which were dominated by a write-down of roughly $10 billion on three investments, most notably $7 billion relating to Arcor. Significantly, Vodafone refused to write down any of its investments in mobile telephony including 3G licences. Net debt was up at $13.5 billion. However, its turnover and operating profit were up significantly, and its worldwide customer base was declared at 95.6 million of whom 90 per cent were active. Hutchison Whampoa then sold its stake down to 2.95 per cent by disposing of 135 million shares.

Conclusions

The case study raises a number of interesting points in seeking to understand how the enormous scale of the reduction in the value of Vodafone transpired. Vodafone's strategy of expansion via acquisition is hard to fault overall in the light of the circumstances of the time, and this explains why, ultimately, the reputation of the Vodafone management has until recently survived the sniping of critics blessed with hindsight. One has only to see how debt piled up at the likes of BT and Deutsche Telekom to appreciate that the extensive use of its own shares to finance expensive takeovers left Vodafone's finances in relatively good shape. However, there were risks involved because a massive share overhang was created and these shares, if released into the market on a large scale, and in close proximity, would obviously depress Vodafone's share price significantly and for some time – as indeed transpired from mid-2001 onwards. The point was nevertheless that most of the shares were not initially expected to be released because investing in Vodafone was seen as a better bet than investing in the companies that Vodafone had taken over, and it was hardly Vodafone's fault that TMT stocks in general suffered such a severe collapse during 2001. However, when the share price fell below £1 in May 2002, the critics became more vocal.

Reasonable risk is one thing, but it is possible to argue that the Vodafone strategy was nevertheless flawed in certain respects for which management could reasonably be held responsible. For example, mobile penetration cannot rise exponentially, and although saturation is seemingly not reached until much higher penetration levels are achieved than was thought to be the case even a few years ago, simply buying new networks was never going to enable Vodafone's revenues to grow rapidly for ever. Clearly, Vodafone was not unaware of this awkward fact of mobile life, and this realisation underpinned the move into 3G and a reliance upon mobile data to pick up where mobile voice could not continue to go. Again, this was strategically sound at the time, but what about the licence fees? The problem here is that although in the case of auctions it was left to bidders to calculate how much to bid, and hence they had to rely upon their commercial judgement for better or worse, they also had in many cases (as did Vodafone in nearly every case) to bear in mind their 2G incumbency and the consequences of failing to get a 3G licence. The fact that they sometimes overbid, especially in the UK and Germany, could therefore be said to be more a case of force of circumstance than of managerial miscalculation. Vodafone certainly did nothing that the other major telcos were unprepared to do, and, significantly, of which the financial markets were unsupportive at the time. When the financial markets withdrew their support, Vodafone, like other carriers, generally bid much less. Significantly, however, Vodafone

Table 6.3 Vodafone and Orange: Strengths and weaknesses

Vodafone	Orange
Strengths	
Global network	Brand recognition
3G licences	Modest debts
Financial muscle	Well managed
Nimble decision-making	
Weaknesses	
Slow growth of profits	No network in USA and Japan
Lack of control in USA and France	Retirement of inspirational CEO
Stock overhangs	MobilCom threat
Different 3G standard in USA	

participated in France, the one market where the licence fee was manifestly overpriced, an action that could only be partly excused by the desire to protect SFR's 2G incumbency.

The recent technological glitches that have held back the introduction of both GPRS and 3G are less to do with carriers than with equipment manufacturers, but there can be no question that a widespread air of pessimism has begun to pervade analysts' reports assessing the prospects for mobile data. Vodafone, unsurprisingly, continues to have faith in its projections, and points out that it is the analysts who are unable to agree what anything is worth while Vodafone's business model has remained unchanged through good times and bad. Given the explosion of demand for mobile voice and subsequently the SMS, Vodafone may be right to plow on regardless of what analysts believe. Nevertheless, with so many other carriers writing off the goodwill[104] from purchases, Vodafone's $120 billion of goodwill in its accounts does appear difficult to justify. Vodafone may well claim that if assets are valued via a discounted cash flow model, then nothing fundamental has happened to alter the numbers and the goodwill figure should stand. In this respect it is increasingly isolated because critics can readily point to the upsurge in competition, the tendency for ARPU to decline and for subscriber acquisition costs (SACs) to rise, regulatory moves against allegedly excessive roaming and fixed-to-mobile call termination charges and delays in the roll-out of new services, and hence it must remain for now a case of 'only time will tell'.

Notes

1 Council of the European Communities (1998) 'Common position (EC) No. 52/98 adopted by the Council on 24 September 1998 with a view to adopting European Parliament and Council Decision No/98/EC, of ... on the co-ordinated introduction of a third-generation mobile and wireless communications system (UMTS) in the Community', *Official Journal of the European Communities*, C333/56, p. 59 and Annex 1; Council of the European Communities (1999) 'Decision No 128/1999/EC of the European Parliament and of the Council of 14 December 1998 on the co-ordinated introduction of a third-generation mobile and wireless communications system (UMTS) in the Community', *Official Journal of the European Communities*, L17/1, p. 4 and Annex 1.

2 Commission of the European Communities (1997) *Communication from the Commission to the Council, the European Parliament, the Economic and Social Committee and the Committee of the Regions: strategy and policy orientations with regard to the further development of mobile and wireless communications (UMTS)*, COM(97) 513 final (Brussels: CEC) p. 6.

3 Commission of the European Communities (1998) *Proposal for a European Parliament and Council Decision on the co-ordinated introduction of mobile and wireless communications (UMTS) in the Community*, COM(1998) 58 final (Brussels: CEC) p. 11.

4 The Economist (1998) *Public Network Europe 1998 Yearbook* (London: The Economist) p. 34.

5 For a concise discussion of the history of GSM see Pelkmans, J. (2001) 'The GSM standard: explaining a success story', *Journal of European Public Policy*, vol. 8(3), pp. 432–53. See G. Darby (1999) 'Bridging wireless and wired networks', *info*, vol. 1(6), for a discussion of much of the technology mentioned in this article. There is also a discussion of technology at www.itu.int/newsarchive/press/releases/1999/99–24 and on that website, at www.ida.gov.sg/Website/IDAConten ... a05f80caf4c825697e0033cdc5?, at www.cellular-news.com/3G and in Federal Communications Commission (2001) *Final report. Spectrum study of the 2500–2690 MHz band*, pp. 9–10 available at www.fcc.gov/3G. FDD is a technique whereby the uplink and downlink are at different frequencies, whereas with TDD they are on the same frequency. For a discussion of the component parts of a mobile network see the Nokia White Paper *Mobile Network Transmission* at www.totaltelecom.

6 The WRC meets every two or three years – the next meeting is in 2003. Its conferences are organised by the International Telecommunication Union, a specialised agency of the UN. Roughly 190 countries are eligible to attend. The WRC's job is to review and, where necessary, revise the Radio Regulations – the international treaty obligations governing the global use of the radio spectrum and of satellite orbits. A report on WRC-2000 can be found in Federal Communications Commission, *op. cit.*, pp. 10–11. The European Commission's role at WRC conferences is to act as a non-voting observer. Its report on

WRC-2000 is *Communication from the Commission to the Council, the European Parliament, the Economic and Social Committee and the Committee of the Regions – results of the World Radiocommunications Conference 2000 (WRC-2000) in the context of radio spectrum policy in the European Community, COM(2000) 811 final* (Brussels: CEC). In preparation for the next WRC meeting the UMTS Forum is pressing for a decision supporting the use of the 2500–2690 MHz band for 3G expansion in Europe because the other two identified bands are currently designated for 2G use and lack spare capacity. It is felt that a strong message of support for this spectrum band will induce other parts of the world to fall into line.

7 Handsets which provide coverage across a number of frequency bands other than those used in the domestic market are not yet needed by the majority of existing and potential customers. When they are needed they will be available. For example, Qualcomm has already developed integrated chipsets that either support both GSM 900/PCN/GPRS and W-CDMA or GSM 800/PCS and cdma2000 1xRTT. Eventually, no doubt, every possible permutation will technically become feasible on a single chipset, albeit at a very high price and with some cost in terms of handset size, but this will be a potential market essentially comprised of well-travelled businessmen who do not want to haul around a lot of separate devices and may not be economic to supply.

8 For details see http://www.3gpp.org. 3GPP is an international standardisation initiative originally founded by ETSI (Europe), ARIB (Japan) and TI (USA). 3GPP should not be confused with 3G3P – the 3G Patent Platform Partnership which is concerned with intellectual property rights – see J. Taaffe, '3G devices – plotting a course for 3G', *Communications Week International*, No. 265, May 2001.

9 Commission of the European Communities (1998) *op. cit.* and 'Amended proposal for a European Parliament and Council Decision on the co-ordinated introduction of mobile and wireless communications in the Community', *Official Journal of the European Communities*, C276/4.

10 Council of the European Communities (1998) *op. cit.*, p. 63.

11 Gruber, H. and Hoenicke, M. (1999) 'The road ahead towards third generation mobile communications', *info*, vol. 1(3), pp. 258–60. A comprehensive, albeit highly technical introduction to GPRS can be found in Ghribi, B. and Logrippo, L. (2000) 'Understanding GPRS: the GSM packet radio service', *Computer Networks*, vol. 34(5), pp. 763–79.

12 MVNOs do not need to buy a licence or build a network. However, to be successful they probably need to have an existing relationship with a customer base that they can exploit, a well-known brand, a distinct set of charges, and their own distribution channels and billing systems.

13 See the section on i-mode in Chapter 2. See also The Economist (2000) 'i-modest success', *The Economist*, 11 March, pp. 97–8. A dissenting view concerning the prospects for i-mode in Europe and the USA is expressed in a report by Tarifica entitled *I-mode in Europe, Entertainment and Content for the European Audience*, published in June 2001, which argues that the technological problems are modest in terms of running i-mode over GPRS and that it will be introduced on a significant scale. It may be noted that i-mode is not sold on the basis of its technological prowess but its ability to provide a wide range of services, and hence DoCoMo is providing a range of 3G handsets to suit the different levels of service sophistication desired by customers.

Meanwhile, WAP version 2.0 was introduced in August 2001 with the capability to support xHTML as well as WML, and DoCoMo announced that it would be attempting to make i-mode/FOMA compatible with WAP 2.0 as they both converged on XML. For a discussion of mark-up languages see UMTS Forum (2001) *3G Portal Study*, Report No. 16 (November) pp. 20–3.

14 See Brisibe, T. (2001) 'GMPCS at the crossroads: the fallout from 3G', *info*, vol. 3(2), pp. 153–8.

15 See Common Market Law Reports (CMLR) (1995) 'Iridium: Globalstar', *CMLR Antitrust Reports* (July), p. 21, and Commission of the European Communities (1997) 'Commission Decision of 18 December 1996 relating to a proceeding under Article 85 of the EC Treaty and Article 53 of the EEA Agreement (Case IV/35.518 – Iridium)', *Official Journal of the European Communities*, L16/87.

16 See The Economist (1998) *Public Network Europe 1998 Mobile Yearbook* (London: The Economist), p. 39 and Wright, D. (1997) 'Obtaining global market access for GMPCS', *Telecommunications Policy*, vol. 21(9/10), pp. 775–83.

17 Very small aperture terminals (VSATs) permit the reliable transmission of data by satellite, using a comparatively small antenna of three to six feet. This can be attached to existing terminal equipment, operating like an aerial modem.

18 See Frieden, R. (1998) 'That pesky last mile. Call termination strategies for mobile satellite systems', *Telecommunications Policy*, vol. 22(2), pp. 133–44.

19 For details see Commission of the European Communities (1995) *Proposal for a European Parliament and Council Decision on an action at a Union level in the field of Satellite Personal Communications Services in the European Union*, COM(95) 529 final (Brussels: CEC). See also *Amended Proposal*, COM(96) 467 final, pp. 17–18, and CMLR, *op. cit.*, pp. 21–2.

20 Financial Times (1999) 'Survey: FT Telecoms, Section 1', *Financial Times*, 9 June, p. VIII.

21 A possible solution is the use of short-range radio links using, for example, the Bluetooth standard to link a handset used indoors to an antenna located on the outside of a building. See Intercai Mondiale (2001) *The Future of Mobile Satellite Services* in the 'White Paper' section of the www.totaltele.com website.

22 See Brisibe, T. (2001) 'GMPCS at the crossroads: The fallout from 3G', *info*, vol. 3(2), pp. 153–8. In March 2001 the $800 million Orbcomm network was sold out of bankruptcy debt-free for $16 million to Advanced Communications Technologies (ACTI) which committed itself to the launch of six replacement satellites in 2003. The target market was identified as checking inanimate objects.

23 Common Market Law Reports (1996) 'Re Iridium (Case IV/35.518)', *CMLR Antitrust Reports* (November), pp. 599–608.

24 See Common Market Law Reports (1997) 'Iridium', *CMLR Antitrust Reports* (February), pp. 187–9 and Commission of the European Communities (1997) 'Commission Decision of 18 December 1996 relating to a proceeding under Article 85 of the EC Treaty and Article 53 of the EEA Agreement (Case IV/35.518 – Iridium)', *Official Journal of the European Communities*, L16/87.

25 For the full story see Price, C. (1999) 'Iridium: Born on a beach but lost in space', *Financial Times*, 20 August, p. 22.

26 See Common Market Law Reports (1995) *op. cit.*, pp. 19–22 and also International Telecommunication Union (1997) *World Telecommunication*

Development Report 1996/97 (Geneva: ITU), Table 3.5 for a list of regional service providers to which were added Toshiba and China Telecom, bringing the total number of service agreements to over 100.

27 See Common Market Law Reports (1997) *op. cit.*

28 See *Financial Times*, 20 May 1999, p. 27.

29 Our discussion is largely confined to networks proposing to provide global coverage. In addition, there are a significant number of operators with a regional footprint, including, in the Asia-Pacific region alone, the Asian Cellular Satellite System (ACeS), Asia-Pacific Mobile Telecommunications (APMT), the Al Thuraya Satellite Telecommunications Co (Thuraya) and Afro-Asian Satellite Communications. Such operators need only own a single geostationary satellite, costing perhaps $500–600 million, and hence should not face the spiralling costs of the likes of Iridium. See Brisibe (2001) *op. cit.*, pp. 155–6 and Foley, T. (2000) 'The only way is up', *Communications Week International*, No. 248, 17 July. A more general review of satellite systems is to be found in Foley, T. (2001) 'Satellites – satellite surge', *Communications Week International*, No. 271, 24 September.

30 See Commission of the European Communities (1997) *Communication from the Commission to the European Parliament and the Council: The World Radiocommunications Conference 1997(WRC-97)*, COM(97) 304 final (Brussels: CEC).

31 See Frieden (1998) *op. cit.*, pp. 136–7 and Foley (2000) *op. cit.*

32 It is estimated that at least 25 per cent of the USA's landmass will be served neither by cable nor DSL in 2010.

33 For example, NorthPoint Communications filed for Chapter 11 protection in January 2001 and Rhythms NetCommunications in August 2001.

34 A slot such as that occupied by DirecTV at 101 degrees West can be expected to yield revenues running to billions of dollars a year.

35 In part because bandwidth cannot be added post-launch.

36 Such as Eutelsat, GE Americom, Intelsat, Loral Skynet and PanAmSat.

37 Consortia building 'Internet-in-the-sky' projects have a vested interest in talking up their prospects. However, Europe Online provides one example of a company that, in October 2000, lost interest in providing internet connections via the Astra satellite, primarily on the grounds that the speed of data transmission slowed rapidly as increasing numbers of subscribers logged on.

38 See www.cellular-news.com/gprs.

39 Ghribi, B. and Logrippo, L. (2000) 'Understanding GPRS: the GSM packet radio service', *Computer Networks*, vol. 34(5), pp. 763–79.

40 According to the Yankee Group Europe (2001) *Mobile Data Pricing: Will European Consumers Pay a Packet?*

41 Analysys (2001) *Pricing GPRS Services*. November. See also *GPRS Roaming: Technical Options and Strategic Implications*. December.

42 In a report published in January 2002 entitled *Learning from i-mode II*, Swedish consultants Northpoint nevertheless make the point that two-thirds of the official content providers do not in fact charge for content, but rather prefer to use their access to i-mode subscribers to direct them to other media interests owned by the providers.

43 For further details see Logica (2001) *Enabling the i-mode Experience beyond the Japanese Frontier*, a White Paper available on the www.totaltele.com website.

So far, the only operational i-mode imitation is an incompatible system in South Korea.

44 This is noted in Standage, T. (2001) 'The Internet, untethered: A survey of the mobile Internet', *The Economist*, 13 October, pp. 5–6.

45 Standage makes the point (*ibid.*, p. 15) that there are some potentially crucial differences between the USA and elsewhere. In particular, whereas the Japanese and Europeans often commute by public transport, leaving them free to play with their mobile phones en route, Americans prefer where possible to go by car. On the other hand, Americans spend far more time hanging around airports where they want to connect up their laptop computers. Furthermore, the American market for mobile connectivity is largely driven by the corporate sector as against the consumer market elsewhere.

46 DoCoMo holds a 15 per cent stake in unlisted KPN Mobile. It announced in November 2001 that it was delaying the creation of a joint venture with KPN Mobile. Instead, it agreed to license a version of i-mode technology covering the period December 2001 to January 2012. E-Plus, the German subsidiary of KPN Telecom and KPN Mobile, is on target for a small-scale launch of i-mode, trading as 'Euro i-mode' over a GPRS network combining cHTML and WAP in a single browser, commencing at the beginning of 2002. However, most analysts expect the technical difficulties inherent in creating such a browser to cause delays. Other operators have tried to introduce a surrogate form of i-mode: TIM, for example, tried inserting a GPRS receiver into a PDC handset but the experiment was a failure. This suggests that if i-mode comes to Europe, it will either come as a direct, authorised copy or not at all.

47 See Taaffe, J. (2000) 'Mobile and satellite: DoCoMo's success fuels debate over WAP', *Communications Week International*, 17 July 2000 (see www.totaltele.com website) and Handford, R. (2000) 'Rival Japanese technology makes a convincing case', *Financial Times*, FT Telecoms Supplement, 20 September, p. XIX. The GSM Association's M-Services initiative is intent upon producing an i-mode equivalent based upon WAP and GPRS, but cannot get participants to agree upon how to configure handsets.

48 For details see http://umts.tkc.at.

49 For details see www.regtp.de/en/umts.

50 For details see www.eett.gr.

51 For the guidelines see www.agcom.it and www.totaltele.com, and for auction details see www.palazzochigi.it and www.comunicazioni.it.

52 For details see www.dep.no/sd and www.npt.no.

53 For details see www.umtsauction.ch.

54 Java is a programming language developed by Sun MicroSystems that runs mini-programmes called applets. It will facilitate location-based services and boost the security of online transactions – see Thompson, V. (2001) 'Platforms – Java blend', www.totaltele.com/roam/view.asp?ArticleID=40491, 1 June, and www.totaltele.com, 27 December 2000. The mobile version of Java, J2ME, faces increasing competition from Qualcomm's binary runtime environment for wireless (BREW) and, in Korea, a local standard called GVM. However, given that BREW is mobile-specific and CDMA-specific it will probably struggle to make inroads into Java's market share, although it was endorsed by Verizon Wireless in February 2002. It should also be borne in mind that the PDC standard (see glossary) used on DoCoMo's 2G network

is faster than GSM, so the initial version of W-CDMA in Japan does not appear to provide an improved speed of connection to i-mode. This, together with the tendency for PDC to provide a relatively long battery charge, at least for now, helps explain the initial adverse user responses in the W-CDMA trials (see 'FT-IT Review', *Financial Times*, 8 August 2001, p. IV).

It should be borne in mind that the issue of harmonisation is not confined to Java. For example, there is a need to achieve data compression if limited memory storage is not to be overwhelmed. DoCoMo's FOMA, for example, utilises (1) MPEG-4 moving image compression technology already developed for other purposes (2) multi-rate transmissions which, to take account of the need for a 3G network to deal with voice, data and images, chooses the best speed and signal for each type of content and (3) rake reception technology which reduces fluctuations in the strength of incoming and outgoing signals that occur as the location of the handset changes – see *NTT DoCoMo Annual Report 2001*, p. 11.

55 For details see www.ida.gov.sg.

56 The detailed discussion and copies of the reports are to be found on www.fcc.gov/3G.

57 For details see www.ursec.gub.uy.

58 For details see http://auction.med.govt.nz.

59 For details see www.aca.gov.au.

60 For an advocacy of auctions compared to other methods of allocation see McMillan, J. (1995) 'Why auction the spectrum?', *Telecommunications Policy*, vol. 19(3), pp. 191–9. However, in Noam, E. (1997) 'Beyond spectrum auctions: Taking the next step to open spectrum access', *Telecommunications Policy*, vol. 21(5), pp. 461–75, Professor Noam argues that whereas auctions as a method of allocation should be seen in a positive light, this does not make them the best way to deal with spectrum. He advocates the use of 'open spectrum access' whereby, using digital equipment for measurement purposes, users 'pay an access fee that is continuously and automatically determined by the demand and supply conditions at the time'. See also Pritchard, W. (1997) 'The communications spectrum: Frequency allocations and auctions', *Acta Astronautica*, vol. 40(208), pp. 447–453 and Bulow, J. and Klemperer, P. (1996) 'Auctions versus negotiations', *American Economic Review'*, vol. 86(1), pp. 180–94.

61 See McMillan, J. (1994) 'Selling spectrum rights', *Journal of Economic Perspectives*, vol. 8(3), p. 147; Guala, F. (2001) 'Building economic machines: The FCC auctions', *Studies in History and Philosophy of Science Part A*, vol. 32(3), p. 455; Scanlon, M. (2001) 'Hiccups in US spectrum auctions', *Telecommunications Policy*, vol. 25(10–11), pp. 689–701; and Milgrom, P. (1998) 'Game theory and the spectrum auctions', *European Economic Review*, vol. 42(3–5), pp. 771–8.

62 See McMillan, *op. cit.*

63 See Guala, *op. cit.* Scanlon, *op. cit.*, conducted an extensive analysis of the early US auctions and concluded that participants were almost certainly aware of the flaws in the auction designs, especially leaving bidders too long a time to exercise default options and rushing to complete the auctions before finalising the rules. He concludes from this that what other countries need to avoid is 'not auctions but political interference which ties the hands

of the auction designers'. He accepts that collusion can occur under auction conditions, but doubts that it has been a material factor and points out that collusion can also be a characteristic of 'beauty contests'. Overall, he believes that the auctions were successful in that they raised a large sum of money, involved relatively little – and increasingly less – defaulting, speeded up the process of licensing and ended up with the spectrum in the hands of those intent upon putting it to its highest-value uses.

64 A justification for the new method is to be found in McAfee, R. and McMillan, J. (1996) 'Analyzing the airwaves auction', *Journal of Economic Perspectives*, vol. 10(1), pp. 159–75. See also Cramton, P. (1998) 'Ascending auctions', *European Economic Review*, vol. 42(3–5), pp. 745–56.

65 See Plott, C. (1997) 'Laboratory experimental testbeds: Application to the PCS auction', *Journal of Economics and Management Strategy*, vol. 6, pp. 605–38 and Cramton, *op. cit.*

66 See Melody, W. (2001) 'Spectrum auctions and efficient resource allocation: Learning from the 3G experience in Europe', *info*, vol. 3(1), p. 8.

67 This illustrates the potential flexibility of auctions. Licences can also, for example, be set aside for minority groups (see New Zealand above) or for entrepreneurial companies (see USA above). It is also possible to cap either the total amount of spectrum to be acquired by a single bidder or to restrict the geographical coverage of a single bidder's licences (see Australia and the USA above). However, it should be noted that there is no reason why similar flexibility cannot be introduced into 'beauty contests'.

68 Melody, *op. cit.*, p. 6.

69 For a detailed discussion of the economics of auctions see the work of Paul Klemperer at www.nuff.ox.ac.uk/economics/people/klemperer. See also Cave, M. and Valletti, T. (2000) 'Are spectrum auctions ruining our grandchildren's future?', *info*, vol. 2(4), pp. 347–50. For various papers containing a detailed analysis of the German auction, contesting the view that the result was in any way 'bizarre', contact wolf@wiwi.hu-berlin.de.

70 The Economist (2000) 'Giddy bidding', *The Economist*, 15 April, p. 36. It must also be noted, however, that continuing to bid when the original upper price limit thought to be prudent has been breached is fostered by the tendency of managers to believe that, since they are better than the management of other bidders, they must be able to make as much profit even if paying a slightly higher price.

71 See, for example, Ure, J. (2001) 'Licensing third-generation mobile: A poisoned chalice?' *info*, vol. 3(1), p. 11. He also poses the question as to whether the industrial markets or the financial markets have judged the issue of auctions correctly since the response of the latter to the structures introduced by the former has been to mark down share prices and credit ratings. See also Noam, *op. cit.*, pp. 465–7.

72 See Financial Times (2000) 'Survey: Telecoms', *Financial Times*, 17 January, p. IV. Subsidies on 2G handsets are being phased out, but with the semiconductor element of a 3G handset costing roughly $150 compared to $50 in a GSM handset, it is difficult to see how subsidies can be avoided when 3G is introduced.

73 Historically, vendor finance covered equipment cost plus some working capital, but such largesse was not forthcoming during 2001, especially in the

light of Lucent's $300 million outstanding loan to One.Tel of Australia which went into administration in June 2001. More recent data are to be found in Financial Times (2001) 'Survey: FT Telecoms', 21 November, pp. 8–9.

74 See, for example, www.oftel.gov.uk/publications/mobile/infrashare0501. According to a June 2001 report by Analysys entitled *Evaluating the Business Case for 3G Network Sharing*, the potential savings would be over five times as high in Sweden compared to Singapore over a ten-year period, with the figure for Germany estimated at 18 per cent – see *Financial Times*, 13 June 2001, p. 27. An extensive analysis of the problems created in trying to co-site 2G and 3G base stations is to be found at www.mfc-net.com in a White Paper published on 3 May 2001 entitled 'The UMTS-GSM co-Siting problem'. According to www.cellular-news.com/article.cgi/5541 of 16 January 2002, the introduction of multi-user detection algorithms devised by Mercury Computer Systems could lead to a reduction of one-third in the number of base stations needed by a 3G network.

75 See, for example, R. Le Maistre, 'Operators: MVNOs – not all Virgins', wysiwyg://9/http://ad.uk.doublclick.ne… m.com/;cat1 = MOBSV; sz = 468 × 60, a report entitled *Secrets of MVNO Success in Western Europe* published by Pyramid Research in January 2002 and www.totaltele.com/newcarrier/view.asp?ArticleID = 39455.

76 See Biddlecombe, E. (2000) 'To the future and back', *Roam* (September) – see www.totaltele.com/roam. See also Taaffe, J. *op. cit.* and 'The future of mobile pre-paid in the 3rd generation of mobile networks' a White Paper of September 2000 at www.totaltele.com

77 At the end of 2001, Samsung SDI announced that it had developed, subject to quality certification, the world's thinnest lithium-ion rechargeable battery for mobile handsets, as light and thin as a polymer battery and capable of storing 355 watt hours.

78 Dundee Investment Research (2000) 'The wireless standards war is over', *Wireless Update*, 6 December – a White Paper available on the www.totaltele.com website.

79 For a full discussion of the issues see Repacholi, M. (2001) 'Health risks from the use of mobile phones', *Toxicology Letters*, vol. 120(1–3), pp. 323–31. The link between radiation and cancers is addressed in Couldwell, C. (2001) 'Mobile health: Risky business', www.totaltele.com/view.asp?ArticleID = 36531 where the point is made that biological damage is associated with ionising radiation whereas mobile phones use non-ionising microwave radiation. Radiation levels are measured via a Specific Absorption Rate (SAR) which registers the absorption of energy by the human body in watts per kilogram. A typical handset registers between 0.5 and 1.0 compared to a maximum safety limit of 2.0. Very recently, in an attempt to provide more conclusive evidence one way or another, the UK government has unveiled 15 research projects, funded half by the private and half by the public sector, which will use human volunteers rather than laboratory experiments that are difficult to replicate and verify. These will form a major plank of a $100 million global project largely co-ordinated by the World Health Organisation. Meanwhile, the USA is struggling to involve public money as there is inevitably scepticism about the results of research funded by the industry – see *Financial Times*, 25 January 2002, p. 9.

80 See, for example, articles in the *Financial Times* (29 December 2000, p. 20; 12 June 2001, p. 14; 17 November 2001, p. 14).

81 For example, the London Borough of Harrow banned a new Orange mast for this reason in March 2001.

82 Commission of the European Communities (2000) *Proposal for a directive of the European Parliament and of the Council on a common regulatory framework for electronic communications and services*, COM(2000) 393 final (Brussels: CEC). This and related matters are discussed in Commission of the European Communities (2001) *Communication from the Commission to the Council, the European Parliament, the Economic and Social Committee and the Committee of the Regions – The introduction of third generation mobile communications in the European Union: state of play and the way forward*, COM(2001) 141 final (Brussels: CEC). See also www.totaltele.com/view.asp?Target = top&ArticleID = 38011.

83 A carrier-class wireless platform capable of delivering 3G-style applications over existing networks via the slashing of bandwidth requirements and simplification of the complexity and power demands on handsets, called eVector Utopia, is close to launch – see www.evectormobile.com.

84 The issue of screen size is being addressed. For example, Microvision has demonstrated a miniature display that uses only three light emitting diodes (LEDs) coupled with a vibrating mirror on a tiny micromechanical chip to project a full-colour image via a small lens which gives the viewer the illusion that s/he is looking at a lap-top screen. One obvious virtue of this is that there is no longer the need to reformat Internet content designed to be viewed on a full-size screen – see www.cellular.news.com/story/5753.

85 A view expounded in, for example, the Yankee Group Europe report entitled *Location Based Services: Positioning the Mobile Internet for Success in Europe?* and the UMTS Forum reports listed below. It may be noted, however, that in the USA mobile operators are being obliged to provide the means to pinpoint the location of people calling for emergency services – the so-called E911 technology – so positioning technology will anyway be embedded and available for other – hopefully profitable – uses. An existing example of successful locational provision is J-Phone's J-Navi service in Japan – see Standage, *op. cit.*, p. 20 – which is provided cheaply as a traffic generator and source of competitive advantage.

86 See www.totaltele.com, 4 December 2000 and 12 December 2000. A discussion of the types of 3G services that might be provided, split into location-based services; infotainment and edutainment; games and remote gambling; travel; educational services; B2C services and many others is to be found in *Enabling UMTS/Third Generation Services and Applications*, 11th Report from the UMTS Forum, chapter 5, available at www.umts-forum.org. See also the 13th Report, *The UMTS Third Generation Market – Phase II: Structuring the Service Revenue Opportunities*; the 14th Report, *Support of Third Generation Services Using UMTS in a Converging Network*; the 16th Report, *A Reference Handbook for Portal Operators, Developers and the Mobile Industry*; and the 17th Report, *The UMTS Third Generation Market Study Update* on the same site.

87 According to DoCoMo, there is no need for a new 'killer application' since the existing range of services plus those already under development will in total contribute sufficient revenue to render 3G economically viable – see www.totaltele.com, 22 May 2001. The absence of a 'killer application' is also

discussed in a 2001 report by IDC entitled *Mobile Data Services and Applications in Western Europe: Forecast and Analysis, 2000–2005*, and in a further report *The Saviour of 3G* issued by Mobile Metrix in January 2002 which argues that there 'are no killer applications per se for 3G, but there are certainly mobile service characteristics' – see www.cellular-news.com/article.cgi/5545 of 16 January 2002. It has been noted elsewhere, possibly with a touch of irony, that 3G's success may ultimately be based on its ability to help 'pass the time' for subscribers. If so, the prototype personal jukebox/music encyclopedia developed by Orange Sverige, Compaq and others should prove to be extremely popular – see www.cellular-news.com/article.cgi/5595 of 23 January 2002. In this regard, it may be noted that handsets may not be marketed as 3G-enabled on the grounds that consumers are indifferent to technology. Rather, handsets will probably be marketed as MP3-enabled, digital camera-enabled and so on.

88 See 'The joy of text', *The Economist*, 15 September 2001, pp. 65–6, and 'Financial Times Survey: FT Telecoms', *Financial Times*, 21 November 2001.

89 The most popular feature mentioned in a December 2001 survey published on behalf of Nokia by the HPI Research Group was indeed SMS texting. It was suggested that its popularity reflected its availability. Those surveyed also expressed considerable interest in being entertained, and in paying for everything via a single bill.

90 Standage, T. (2001) 'The Internet, untethered: A survey of the mobile Internet', *The Economist*, 13 October, p. 15. He makes the arguably crucial point that it costs money to send information whatever the chosen mechanism, and that, for example, even an expensive international text message may seem preferable to a postcard. However, while texting may pander successfully to a world that grows increasingly impatient with delays in data carriage, it also helps instil an attitude in mobile device users that information should appear almost instantly, which can cause a degree of disillusion.

91 See footnote 23.

92 Very recently, some relatively optimistic views have been expressed. For example, consultants at Spectrum Strategy have made the point that although the cost of 3G networks has indeed been extremely, if not excessively, high in some cases – perhaps $10 billion overall per network in the UK – the crucial point is that revenues from the growth of data services, whether delivered over 2G, 2.5G or 3G networks, will more than cover the outgoings and hence there is no real possibility of 3G operators going bankrupt – see www.totaltele.com/vprint.asp?txtID = 41913, 16 July 2001. The view taken by Ovum in its September 2001 report, entitled *3G Survival Strategies: Build, Buy or Share*, is that those companies that avoid being first-movers should prosper in the medium term.

93 W-LAN standards are approved by the Institute of Electrical and Electronics Engineers (IEEE). 802.11a operates in the 5 GHz band at 54 Mbps, but is incompatible with 802.11b which operates at 2.4 GHz at 11 Mbps. In November 2001, the IEEE tentatively approved 802.11g which is compatible with 802.11b but operates at 54 Mbps. 802.11g was held back by disputes between LAN chipmakers Intersil and Texas Instruments, but should be up and running by the end of 2002. Although 802.11b is not designed for outdoor use beyond 50 yards or so, companies such as Proxim have used modified roof-top aerials to achieve much longer ranges. Although Wi-Fi is

officially too power-hungry for small battery-powered devices, it is beginning to be built into handheld computers and thereby encroaching upon the realm of Bluetooth. A single Bluetooth 'master' can communicate with up to seven 'slaves' thereby creating a so-called 'piconet'. For a discussion of Bluetooth see, for example, *The Economist Technology Quarterly*, December 2000, pp. 14–20 and articles in *Roam*, September 2000 and November/December 2000. The fact that Windows XP has been configured to support 802.11b rather than Bluetooth has not helped its prospects – see www.totaltele.com/view.asp?ArticleID = 40573. If Wi-Fi and Bluetooth are used in proximity their slightly different modes of operation can cause the Wi-Fi signal to degrade. It must also be borne in mind that the 2.4 GHz band not only has to accommodate W-LANs and Bluetooth, but the likes of microwave ovens and cordless phones. Nevertheless, Bluetooth's prospects are lauded in certain quarters on the grounds that it is more than simply a direct alternative to a W-LAN – see www.totaltele.com/v.print.asp?textID = 41434 and 42418. At the end of October 2001, Manchester in the UK claimed to be the first city in the world to have launched an urban wireless network connected via Bluetooth. At the end of the day, Wi-Fi scores highly on interoperability, but Bluetooth chips are becoming so cheap – $5 apiece is forecast for the end of 2003 – that they will probably be inserted in handheld devices irrespective of whether they work properly, and a separate chip for each major equipment manufacturer may be inserted in laptops to promote interoperability without much affecting their cost.

94 For a summary of progress see www.totaltele.com, 12 June 2001 and 17 August 2001. See also Standage, *op. cit.*, pp. 12, 15. While the use of unlicensed spectrum for private purposes is regarded as acceptable, annexing it for public applications such as LANs and charging for services provided may well instigate regulatory interference – for example, it is illegal to use W-LAN frequencies for profit in the UK – possibly prompted by complaints from 3G licensees. Problems of congestion may anyway induce self-imposed limits on use of the spectrum. So far, companies seeking to create viable businesses based on the provision of W-LAN services have not been successful. For example, Metricom was forced to file for Chapter 11 bankruptcy in June 2001 with debts of $1 billion – see www.totaltele.com/v.print.asp?txtID = 42602. For a variety of views on the future of W-LANS see the 'Current Forum' dedicated to 'Broadband: WLANS or UMTS?' on the totaltele.com website. See also www.etinium.net.

95 See www.totaltele.com, 19 March 2001. The Korean government has also announced the creation of a taskforce to study the development of 4G, and agreed to implement in conjunction with the Japanese government a trial 4G satellite communications service during the World Cup finals which are being jointly hosted by the two countries. Most of the development work is concentrating upon moving to 3.5G via the introduction of high-speed downlink packet access (HSDPA) which is expected to send data from base station to handset at 10 Mbps with a view to moving to 20 Mbps in 2001 – officially 4G.

96 The first half of 2001 witnessed a series of profit warnings by major equipment manufacturers such as Ericsson and Lucent, leading to collapsed share prices. There are signs that manufacturers will either become involved in take-over bids – for example, Alcatel launched a take-over bid for Lucent at the end of May, although it immediately collapsed due to arguments over,

inter alia, issues of control – or joint ventures such as those already created between Sony and Ericsson and Toshiba and Siemens. Japanese manufacturers' experience with i-mode makes them attractive potential partners for European/US equivalents, but the history of joint ventures in the telecommunications sector makes depressing reading – see, for example, Harney, A. and Nakamoto, M. (2001) 'The sun also rises on Japan's 3G mobile push into Europe', *Financial Times*, 17 May, p. 31.

97 This point is made forcefully by Emerson, B. (2000) in *Roam* (November/December) – see www.totaltele.com/roam. He argues that one should 'forget 2 Mbps, even if you're the only user in that cell and the wind is blowing in the right direction'. He claims that GPRS will only operate at up to 50 Kbps and that for 3G even the official specifications only refer to speeds of 144 Kbps.

98 See www.totaltele.com, 21 February 2001 and 22 February 2001. At the components level, Intel announced in May 2001 that it had created a 'wireless-internet-on-a-chip' providing much enhanced battery life and processing power compared to other microchips used in mobile devices. Test chips are expected to become available during 2002 – see www.totaltele.com, 17 May 2001.

99 There are already some signs of progress. For example, in June 2001, Nokia and Sonera announced that they had taken a major step forward in enabling operators to provide customers with different Quality of Service (QoS) levels using Sonera's GPRS Roaming Exchange (GRX) in Nokia's trial UMTS platform. Both the ability to roam freely and to charge according to QoS are viewed as crucial issues to resolve before 3G services commence.

100 Unsurprisingly, network operators want to see a return on their investment as quickly as possible and are pressing content providers for a minimum of 40 per cent of total revenues generated, which is considerably more than the latter are willing to pay over – it may be noted that DoComo is much less demanding in relation to i-mode since it charges only 9 per cent of revenue and relies upon gross turnover rather than margin to generate acceptable returns.

101 The euro (€) fell fairly steadily against the dollar throughout much of the period under discussion, falling as low as $0.83. Conversion into dollars has been undetaken as appropriate at the rate ruling at the time specified.

102 Earnings before interest, tax, depreciation and amortisation.

103 In mid-1999, there were roughly 15 billion shares issued. This then shot up to 30 billion and, early in 2000, to 60 billion. By mid-2001, there were almost 70 billion issued – hence, at £1.50 or so apiece, yielding a total value of roughly £100 billion ($145 billion) for the entire company.

104 The value of takeover bids invariably exceeds the estimated value of the assets acquired, sometimes by a substantial margin. However, the size of the difference is fluid, and the purchaser obviously has to believe in principle that the target's value will rise substantially once it is under better management, and hence that the goodwill element of the purchase will gradually disappear over time without any further action needing to be taken. If it clearly has not disappeared, it will eventually need to be 'written off' in the accounts with damaging effects upon profitability. The way in which goodwill has to be accounted for varies by country.

Index